MMA, A Computer Code for Multi-Model Analysis

Constructed using the JUPITER API
    JUPITER: Joint Universal Parameter IdenTification and Evaluation of Reliability
    API: Application Programming Interface

By Eileen P. Poeter and Mary C. Hill

Prepared in cooperation with the
U.S. Environmental Protection Agency, U.S. Department of Energy, and
International Ground Water Modeling Center, Colorado School of Mines

Techniques and Methods 6-E3

U.S. Department of the Interior
U.S. Geological Survey

U.S. DEPARTMENT OF THE INTERIOR
DIRK KEMPTHORNE, Secretary

U.S. GEOLOGICAL SURVEY
Mark D. Myers, Director

U.S. Geological Survey, Boulder, Colorado: 2007

For product and ordering information:
World Wide Web: http://www.usgs.gov/pubprod
Telephone: 1-999-ASK-USGS

For more information on the USGS – the Federal source for science about the Earth,
its natural and living resources, natural hazards, and the environment:
World Wide Web: http://www.usgs.gov/pubprod
Telephone: 1-888-ASK-USGS

Suggested citation:
Poeter, Eileen P., and Mary C. Hill, MMA, A computer code for Multi-Model Analysis: U.S. Geological Survey Techniques and Methods 6-E3, 113 p.

Preface

The Multi-Model Analysis (MMA) computer code is designed to evaluate many alternative models of a given system, called multiple models in this work. It can be used to rank the models and calculate posterior model probabilities. The probabilities are used to calculate model-averaged quantities that account for the variability evident in the alternative models. The model-averaged quantities can include parameter estimates, predictions, and measures of parameter and prediction uncertainty. Calibration of all models needs to be completed before application of MMA.

MMA operates by reading files produced by models developed and calibrated to represent a single system. The models all need to use the same observations with the same weighting. The files needed by MMA can be produced by UCODE_2005 and associated codes (Poeter and others, 2005) run in the Sensitivity-Analysis or Parameter-Estimation mode. The name and structure of the files are simple and follow the conventions of JUPITER API data-exchange files. Thus, it is likely that MMA also will be able to use results from codes other than UCODE_2005. Although the examples presented in this work are from the field of ground-water modeling, MMA (and UCODE_2005) can be used to evaluate results from models of nearly any type of system.

The documentation presented in this report describes the methods used and the input and output files. Together, this report, Hill and Tiedeman (2007), and Burnham and Anderson (2002) serve to document the computer code described in this work. Additional information on the methods is provided in Burnham and Anderson (2004), Ye and others (2004, 2005), and Poeter and Anderson (2005).

The performance of MMA has been tested in a variety of applications. Future applications, however, might reveal errors that were not detected in the test simulations. Users are requested to notify the originating office of any errors found in the report or the computer program. Updates might occasionally be made to both the report and the computer program. Users can check for updates on the World Wide Web at URL http://water.usgs.gov/software/ground_water.html/ or http://water.usgs.gov/software/general.html/.

Contents

Figures

Tables

MMA, A Computer Code for Multi-Model Analysis

By Eileen P. Poeter[1] and Mary C. Hill[2]

Abstract

This report documents the Multi-Model Analysis (MMA) computer code. MMA can be used to evaluate results from alternative models of a single system using the same set of observations for all models. As long as the observations, the observation weighting, and system being represented are the same, the models can differ in nearly any way imaginable. For example, they may include different processes, different simulation software, different temporal definitions (for example, steady-state and transient models could be considered), and so on. The multiple models need to be calibrated by nonlinear regression. Calibration of the individual models needs to be completed before application of MMA.

MMA can be used to rank models and calculate posterior model probabilities. These can be used to

(1) determine the relative importance of the characteristics embodied in the alternative models,

(2) calculate model-averaged parameter estimates and predictions, and

(3) quantify the uncertainty of parameter estimates and predictions in a way that integrates the variations represented by the alternative models.

There is a lack of consensus on what model analysis methods are best, so MMA provides four default methods. Two are based on Kullback-Leibler information, and use the AIC (Akaike Information Criterion) or AICc (second-order-bias-corrected AIC) model discrimination criteria. The other two default methods are the BIC (Bayesian Information Criterion) and the KIC (Kashyap Information Criterion) model discrimination criteria. Use of the KIC criterion is equivalent to using the maximum-likelihood Bayesian model averaging (MLBMA) method. AIC, AICc, and BIC can be derived from Frequentist or Bayesian arguments. The default methods based on Kullback-Leibler information have a number of theoretical advantages, including that they tend to favor more complicated models as more data become available than do the other methods, which makes sense in many situations.

Many applications of MMA will be well served by the default methods provided. To use the default methods, the only required input for MMA is a list of directories where the files for the alternate models are located.

Evaluation and development of model-analysis methods are active areas of research. To facilitate exploration and innovation, MMA allows the user broad discretion to define alternatives to the default procedures. For example, MMA allows the user to (a) rank models based on model criteria defined using a wide range of provided and user-defined statistics in addition to the default AIC, AICc, BIC, and KIC criteria, (b) create their own criteria using model measures available from the

---

[1] International Ground Water Modeling Center and the Colorado School of Mines, Golden, Colorado, USA.
[2] U.S. Geological Survey, Boulder, Colorado, USA.

code, and (c) define how each model criterion is used to calculate related posterior model probabilities.

The default model criteria rate models are based on model fit to observations, the number of observations and estimated parameters, and, for KIC, the Fisher information matrix. In addition, MMA allows the analysis to include an evaluation of estimated parameter values. This is accomplished by allowing the user to define unreasonable estimated parameter values or relative estimated parameter values. An example of the latter is that it may be expected that one parameter value will be less than another, as might be the case if two parameters represented the hydraulic conductivity of distinct materials such as fine and coarse sand. Models with parameter values that violate the user-defined conditions are excluded from further consideration by MMA.

Ground-water models are used as examples in this report, but MMA can be used to evaluate any set of models for which the required files have been produced.

MMA needs to read files from a separate directory for each alternative model considered. The needed files are produced when using the Sensitivity-Analysis or Parameter-Estimation mode of UCODE_2005, or, possibly, the equivalent capability of another program.

MMA is constructed using modules and conventions for data-exchange files from the JUPITER API, and is intended for use on any computer operating system. MMA consists of algorithms programmed in Fortran90, which efficiently performs numerical calculations.

Chapter 1: Introduction

When simulating most natural systems, commonly there are alternative plausible models. For example, alternative models of a ground-water system may be developed due to uncertainty associated with the following:

(1) The structure and character of boundary conditions.

(2) Relevant processes (which might require models to use different simulation software).

(3) The spatial and temporal distribution of system characteristics such as hydraulic conductivity, recharge, reaction coefficients, and so on, including alternatives based on different ideas about the deposition and deformation of geologic materials.

(4) The inclusion or exclusion of transients associated with, for example, pumping rates, source concentrations, recharge, and so on. The importance of their variation on annual, seasonal, monthly, daily, or other temporal scale, might be of concern.

Alternative ideas about how a system is best represented are often controversial because the model differences result from strongly held beliefs about the system or different models produce substantially different practical, financial, and(or) regulatory consequences. The multi-model analysis methods presented in this work can focus such controversy on a constructive process of hypothesis testing, modeling, and data collection that is more likely to lead to consensus and wise decisions.

The first step is to develop the alternative models, which can be accomplished using any method. There is general agreement that considerable mental effort, training, and experience, are required to define a set of meaningful alternative models and that the effort is important to the evaluation of natural systems (Bredehoeft, 2003; Neuman and Wierenga, 2003). All models considered need to be evaluated using the same set of observations and observation weighting. The models need to be calibrated before using MMA.

Model development typically requires many model runs regardless of how it is accomplished. The ability to develop many alternative models which may each require substantial computational effort is possible because of the advent of high-speed computing and robust models, solvers, inversion algorithms, and sophisticated graphical user interfaces. These are some of the reasons for a recent increased interest in multi-model methods as applied to models of complex natural systems.

A variety of criteria have been suggested for multi-model analysis (Carrera and Neuman, 1986; Neuman and Wierenga, 2003; Ye and others, 2004, 2005; Poeter and Anderson, 2005). Thus, MMA defines a set of default criteria instead of a single criterion, and it also provides flexible mechanisms with which other criteria can be defined. For each application of MMA, prior model probabilities can be defined or not. Posterior model probabilities can be calculated with each criteria using a single default method (for which there is wide support in the literature), or the user can specify an alternative.

Chapter 1: Introduction

The four default model criteria provided by MMA are: (1) the Akaike information criterion (AIC), (2) the second-order-bias corrected AIC (AICc), (3) the Bayesian information criterion (BIC), and (4) Kahyap's information criteria (KIC). AICc approaches AIC as the number of observations increases. Default criteria (1) to (3) are based only on model fit to observations and the number of estimated parameters. The KIC default method also requires the Fisher information matrix, which is calculated using sensitivities (the derivative of simulated equivalents with respect to parameter values). UCODE_2005 (Poeter and others, 2005) can be used to calculate the needed results. Other programs such as PEST (Doherty, 2004) or Ostrich (Matott, 2005) also could be used, but the results would need to be written to data-exchange files for use by MMA. While MMA was developed primarily to use the results from models calibrated using single objective-function nonlinear regression methods, other methods could be used if the data-exchange files needed by MMA are produced. It is also likely that many of the default model criteria could be used with the alternative regression methods, such as multi-objective function methods, but this has not been investigated thoroughly.

Measures of uncertainty that reflect the existence of multiple plausible models can be calculated with MMA using the posterior model probabilities. Often, measures of predictive uncertainty evaluated using a range of alternative conceptual models are larger than measures obtained based on the results of any one model. Indeed, confidence intervals on predictions from some models may not include predictions from other models. This raises the question of whether to select the best model and use those predictions and confidence intervals for decision and design, or to consider some or all of the models and calculate model-averaged predictions and intervals. This issue is discussed in chapter 2 of this report.

Purpose and Scope

This report documents MMA, a multi-model analysis code. Readers of this report may come from many backgrounds, because MMA can be used with process models from any discipline. Different fields tend to have their own problems related to model evaluation and their own literature addressing these problems. The reader is encouraged to become familiar with these resources.

This report begins with an overview of how MMA calculates the individual model criteria and how these criteria are used to calculate posterior model probabilities and model-averaged values. The remainder of the report describes, in detail, how to run MMA, construct input files, and use the MMA output files. Appendix 1 describes the connection between MMA and the JUPITER API. Appendix 2 includes an example application with MMA input and output files for a simple problem. Appendix 3 contains information about the distributed files, including source code files. Appendix 4 discusses discrepancies between the equation for the KIC criterion presented in this and previous works.

Files for MMA are available on the World Wide Web at URLs
http://water.usgs.gov/software/ground_water.html/ and
http://water.usgs.gov/software/general.html/.

Expertise of the authors is primarily in the simulation of ground-water systems, so examples in this report come from this field. However, models of nearly any type of system can be evaluated.

The first requirement for using MMA is to be knowledgeable about the process model(s) being investigated. Without such knowledge any statistical technique is likely to be used poorly. The second requirement is that the user has some knowledge about basic statistics and the application of nonlinear regression. For example, it is assumed that the reader is familiar with the terms "standard deviation, variance, correlation, sensitivity, optimal parameter values, residuals, and probability." Readers who are unfamiliar with these terms are encouraged to review a basic statistics book, such as Helsel and Hirsch (2002) and other references and applications cited in Hill (1998) and Hill and Tiedeman (2007). The terms prior model probability and posterior model probability are described and used extensively in this document; the reader is not expected to have previous experience with these terms. Discussion of the advantages of AICc is provided by Poeter and Anderson (2005) and Burnham and Anderson (2002, 2004). Discussion of the advantages of KIC is provided by Ye and others (2004, 2005).

Acknowledgments

The authors would like to acknowledge John Doherty, of Watermark Computing and the University of Queensland, whose Equation module in the JUPITER API (Banta and others, 2006) is used in MMA. We would also like to acknowledge Laura Foglia who, as a Ph.D. student at ETH in Zurich, Switzerland, used early versions of MMA and provided useful suggestions. We also appreciate the unusually extensive review of this document provided by Professor Noel Merrick of the University of Technology in Sydney, Australia. Dr. Shawn Matott, now at the U.S. Environmental Protection Agency in Athens, Georgia, USA, Professor Ty Ferre, of the University of Arizona, and Professor Ming Ye, of Florida State University, also provided very useful review comments.

Finally, the first author would like to thank the National Ground Water Association, which provided the opportunity for her to discuss many of the ideas in this report in 60 presentations given in North America, Europe, Australia, China, Thailand, and South Africa as part of the 2006 Henry Darcy Distinguished Lecture Series.

Chapter 2: Methods for Multi-Model Analysis

This chapter presents an overview of how MMA operates, some thoughts about developing alternative models, a list of circumstances for which proposed models may be omitted from consideration, and some ideas on using individual-model and model-averaged results. Then, the analyses provided as defaults by MMA are described and discussed. Finally, alternative methods accessible through MMA are discussed.

Overview

MMA uses data-exchange files generated by running alternative models of one system using the same set of observations. Specifically, MMA was developed to do the following:

(1) Gather results produced by the models. For the default analyses, the results needed by MMA are, for example, (a) the sum of squared weighted residuals, (b) the number of estimated parameters, and (c) the natural log of the determinant of the Fisher information matrix, which requires observation sensitivities and weighting in its calculation. Depending of the options selected, MMA may also need (d) optimized parameter values, and(or) (e) model predictions. Generally, the results from each model are located in a different directory, and the directories are listed in the MMA Model_Paths input block described in chapter 5. The output files in those directories need to be named and constructed as required for MMA to find and read them.

(2) Conduct a test that indicates that the observations and their weighting are the same in all models. The values of the weights and observations are not checked directly. Instead, the names assigned to the observations are compared for each model. For example, in UCODE_2005, the names for the observations are assigned using keyword ObsName of the Observation_Data input block (Poeter and others, 2005, p. 83). For each model, the names are printed in a data-exchange file with filename extension _os, and this file is read by MMA.

(3) Models are ignored under selected circumstances, as discussed in the section below entitled "Omitting selected models from the analysis." The default is described there.

(4) Compute measures of model quality. This can include a wide range of measures, including formal model discrimination criteria, statistics from graphical analysis of residuals, among others.

(5) Define measures of model quality to be used as model criteria. Four default criteria or any number of user-defined model criteria can be used.

(6) Use the criteria to calculate posterior model probabilities. Use evidence ratios and inverted evidence ratios to evaluate the relative probability of the different models. Potentially eliminate unlikely models from further consideration.

(7) For each model criterion, use the posterior model probabilities to rank the models, with the rank of 1 being assigned to the best (most probable) model. If the default method is used to calculate posterior model probabilities, the best model will have the smallest value of the criterion. If user-defined equations are used, that may not be the case.

(8) Use the posterior model probabilities to calculate model-averaged predictions and(or) parameter values and their associated model-averaged confidence intervals.

(9) Print the measures of model quality for all evaluated models, the model rankings, and the model-averaged parameter values and(or) predictions and their confidence intervals.

Steps 4 through 8 can be accomplished using default methods or through user-defined methods specified using statistics and equation capabilities provided by MMA. When using the default methods, MMA makes these calculations for each of four model criteria using one method of calculating posterior model probabilities, as described in this chapter. Other methods can be achieved by specifying an equation to define the model criterion (involving any combination of the model measures) and, possibly, an alternative equation to use those values to determine a posterior model probability for each model.

Multiple Working Hypotheses

To use MMA, multiple models of a system first need to be developed. While a complete discussion of generating multiple models is necessarily specific to the type of system involved, a few general comments are presented here.

Generating multiple meaningful working hypotheses and the resulting multiple models of a given system requires substantial understanding of the system and the data. The multiple hypotheses can be derived from deterministic arguments, such as alternative theories about depositional environments of the sediments or deformation of rocks that make up a ground-water system. They also can be derived from stochastic arguments, such as generating multiple zonations using indicator kriging or pilot point distributions using random sets of locations. Parameterization and optimization techniques such as constrained minimization and super parameters obtained through singular decomposition methods can be useful. These and other methods of generating alternative models and general ideas about alternative models are discussed by many authors, including Bredehoft (2003), Doherty (2003), Franssen and others (2003), Tonkin and Doherty (2005), Moore and Doherty (2005, 2006), Hill (2006), Hill and Tiedeman (2007), and Hunt and others (2007). The number of hypotheses might range from a few to a few dozen or many more. Deterministic methods tend to produce fewer hypotheses, while stochastic processes tend to produce more.

Given a set of data, hypotheses, and models, the first concern is model selection (also called model discrimination). Of concern is identifying which hypothesis or hypotheses clearly explain the observations. If one model is obviously superior, it may be valid to report predictions from a single model, though it may be advantageous to report uncertainty measures that reflect other models.

If no one model is obviously superior, the question is what to do with the multiple valid alternative models. The next two paragraphs comment on this issue assuming that the primary purpose of the model is to produce predictions. Extension to models developed for other purposes such as estimating parameter values is straightforward. Parameters might be of interest, for example, if they are defined to represent the mass-loading or location of a contaminant source.

Given multiple plausible models, it can be useful to report results from individual models, including statistics that reflect model fit and parsimony, and predictions and confidence intervals on predictions. However, an analysis that stops with such a presentation is likely to be cumbersome and confusing if there are many predictions and(or) many models to consider.

In some circumstances, a more useful analysis can be achieved by including model-averaged predictions and confidence intervals that reflect the multiple models considered. The contribution of each model to the model-averaged results is determined by the posterior model probability, as discussed later in this chapter in the section "Multi-Model Inference."

Omitting Selected Models from the Analysis

Models are omitted from the analysis if any of the following three things occur.

(a)   Model regression did not converge. This is communicated to MMA through a data-exchange file with filename extension _dm produced for each model by UCODE_2005 or other calibration program. Models that did not converge can be included in MMA by rerunning them in Sensitivity-Analysis mode for UCODE_2005, or something similar when using another program. This might be desired if the model was thought to be a good representation of the system despite the inability of the regression to converge. In addition, Sensitivity-Analysis mode is needed if parameters not estimated by the regression are to be included in an analysis of uncertainty. For more information about this option, see the section "Parameters for which the values are not estimated by the regression" at the end of this chapter.

(b)   Estimated parameter values are unreasonable, where unreasonable is defined by the user in the MMA Param_Eqns input block described in chapter 5. The default is that no tests of the parameter values are conducted (that is, all parameter values are treated as being reasonable). Additional comments are provided in the section "Eliminating models based on analysis of estimated parameter values" later in this chapter.

(c)   Observations were omitted because of numerical considerations (see additional comments about this in chapter 3 under the definition of the variable NOBS).

The variable R is used in this report to represent the number of models that survive these tests and are analyzed using MMA.

Using the Results of Multi-Model Analysis

If one model is clearly superior to the rest, it is reasonable to use that model for prediction. However, even in that circumstance, it can be advantageous to evaluate prediction error using a larger set of candidate models. If one model is not clearly superior, then it may be reasonable to use a model-averaged prediction. Hill and Tiedeman (2007) suggest that a more representative model displays the following characteristics.

(a)   No dominant spatial or temporal pattern in the weighted residuals or, if patterns exist, they are consistent with expected correlations of the weighted residuals produced by the fitting process of the calibration method.

(b)   Reasonable estimated parameter values and reasonable relative estimated parameter values (for example, in a ground-water model, material known to be gravel is expected to have higher hydraulic conductivity than material known to be silt).

(c)   In general, given similar model fit to observations, it is argues that simpler models are likely to have better predictive capabilities than complex models.

If reasonable alternative models yield substantially different results for the prediction of interest such that a reasonable decision can not be made, additional data collection may be needed to identify likely and unlikely predictions. Tools for evaluating useful data for collection include sensitivity analysis and other methods as described by Wagner and Harvey (1997), Minsker (2003), Tiedeman and others (2004), Hill and Tiedeman (2007), and Tonkin and others (in press), among others.

To understand the model-averaged results, it can be useful to consider in detail the most likely individual models. It is generally these models that dominate model-averaged results. Helpful results include fit-independent sensitivity analysis (using, for example, dimensionless-, composite-, and prediction-scaled sensitivities, parameter correlation coefficients, leverage statistics, the observation-prediction statistic OPR, and the parameter-prediction statistic PPR). Also, influence statistics such as DFBETAS and Cook's D can be useful. These methods can be used to identify observations that are important to the model-averaged estimated parameters, predictions, and confidence intervals, and parameters that are important to the predictions and confidence intervals. For more information, see Hill (1998), Poeter and others (2005), and Hill and Tiedeman (2007). All of these statistics are produced by UCODE_2005 and programs distributed with it. An example is provided in Appendix 2.

The default methods used for multi-model analysis are discussed next. That is followed by a discussion of alternate methods that can be accessed through MMA.

MMA Defaults for Multi-Model Analysis

MMA provides four default methods for model analyses, which reflects the lack of consensus about the superiority of any one method. The analyses differ in the criterion used. The four analyses are presented in two pairs for which one criterion is the asymptotic limit of the other. The first pair includes the AIC and AICc criteria; the second pair includes the BIC and KIC criteria.

Alternatives to the default model criteria can be defined using the Analyses input block described in chapter 5. If alternative model criteria are defined, they are calculated instead of, not in addition to, the default criteria. To obtain a combination of AIC, AICc, BIC and(or) KIC and other criteria, they would all need to be defined in the Analyses input block.

Multi-Model Analysis Methods that are Estimates of Relative Kullback-Leibler (K-L) Information

Multi-model analyses derived from Kullback-Leibler (K-L) information are based on the concept that models are approximations (that is, there are no true models of complex systems). They tend to select models with more parameters as the number of observations increases, which is consistent with the idea that smaller effects can be identified in complex systems as the number of observations increase.

Kullback-Leibler (K-L) information is based on a coherent theory of model selection that has been developed over the past 30 years. It has been the subject of text books (for example, Linhart and Zucchini, 1986; McQuarrie and Tsai 1998; Burnham and Anderson, 2002), research monographs (for example, Sakamoto and others 1986), and hundreds of journal papers (for example, deLeeuw 1992).

The starting point is K-L information, $I(f,g)$ (Kullback and Leibler 1951). This is interpreted as the information, I, lost when full truth, f, is approximated by a model, g. Given a set of candidate models g_i, one might compute K-L information for each of R models and select the one that minimizes information loss – that is, minimizes $I(f,g)$. This is a compelling approach. However, for models of natural systems (for example, ground-water systems), K-L information cannot be computed because the true model and the optimal effective parameters (for example, hydraulic conductivities, boundary heads and fluxes) are not known (Anderson 2003). A workable approach is to consider the change in K-L information between pairs of alternative models, which forms the basis of the approach described here.

AIC and AICc Criteria

The AIC and AICc criteria are estimators of twice the expected K-L information loss. Akaike (1973, 1974) developed a way to estimate expected K-L information, based on a bias-corrected maximized log-likelihood value; the resulting statistic has two terms and is commonly referred to as AIC.

$$\text{AIC} = n \ln(\sigma^2) + 2\,k \qquad (2.1)$$

The variables are defined after equation 2.2a.

Better approximations to the bias are presented by Sugiura (1978) and Hurvich and Tsai (1989, 1994), and result in an equation referred to as AICc which is calculated as

$$\text{AICc} = n \ln(\sigma^2) + 2\,k + \left(\frac{2\,k\,(k+1)}{n-k-1} \right), \qquad (2.2a)$$

where:

n is the number of observations plus, in some circumstance, the number of prior information equations on the parameters (See the section entitled "Including prior information on parameters" later in this chapter and the discussion for NOBS in chapter 3 for additional information);

k is the number of parameters, and equals NPE+1, as discussed in the following text;

NPE   equals the number of process model parameters;

ln   is log base e; and

σ^2 is the residual variance and is estimated in these equations based on maximum-likelihood theory as

$$\sigma^2 \approx s_{ML}{}^2 = \mathrm{SWSR}/n. \tag{2.2b}$$

SWSR is the sum of weighted squared residuals objective function, which, given a diagonal weight matrix on the observations, is calculated as

$$\mathrm{SWSR} = S(b) = \sum_{i=1}^{n} \omega_i \left[y_i - y'_i(\underline{b}) \right]^2 \tag{2.2c}$$

where,

$\underline{b}$ is a vector containing the values of NPE parameters;

ω_i is the weight for the ith observation and applies when the observation errors are independent (use of a full weight matrix is discussed in the following text);

y_i is the value of the ith observation or, sometimes, the ith prior information value; and

$y'_i(\underline{b})$ is the simulated equivalent that is compared to the ith observation.

The variables k and ω_i are discussed further in the next three paragraphs.

Defining k as NPE+1 is consistent with assuming a normal distribution, as follows. When errors are normally distributed, the only parameter in the probability distribution that needs to be estimated is σ^2 because the mean of the errors is assumed to be zero. Parameters of the probability distribution need to be estimated when using the likelihood function, $L(\underline{b}, s^2 | \underline{y}, g)$. In words, $L(\underline{b}, s^2 | \underline{y}, g)$ represents the likelihood of the estimates ($\underline{b}$), of the true, unknown parameter values (β) and the estimate (s^2) of the true, unknown error variance (σ^2), given the observation data ($\underline{y}$) and the model (g). A likelihood can be calculated for any set of model parameter values and estimated variance, and is equivalent to the probability $P(\underline{y}, | \underline{b}, s^2, g)$. Given that the variance is estimated along with the parameters of the process model, the total number of parameters equals NPE+1.

If the observation errors are correlated, equation 2.2c is expressed using a weight matrix, $\underline{\omega}$, as discussed by Hill and Tiedeman (2007, p. 34–35, 298). Observation weights are not to be confused with posterior model probabilities, which are sometimes called model weights. Posterior model probabilities are discussed later in this chapter.

Equation 2.2a is the result of a precise mathematical derivation. The second term accounts for first-order bias and the third term accounts for second-order bias resulting from a small number of observations. The third term depends on the assumed distribution of what, in this report, are called true errors (called residuals by Burnham and Anderson, 2002, p. 6). Equation 2.2a is the result when the true errors are normally distributed. Accounting for second-order bias is important when $n/k < 40$, which is typical of many models.

AIC currently is more commonly used than AICc. AICc and AIC tend to select the same model when n/k is large. AICc needs to be used if $n/k < 40$ for any model considered. (Burnham and Anderson, 2002, p. 66).

Delta Values

Generally, model criterion values themselves are not meaningful. Instead, the differences between model criterion values are used to analyze alternative models. The differences are called delta (Δ_i) values and, for a given set of models, are calculated relative to the model with the smallest criterion value. Thus, using AICc as an example, the delta values are calculated as

$$\Delta_i = AICc_i - AICc_{min} \tag{2.3}$$

for each model, i, in a set of R models being analyzed, where $AICc_{min}$ is the minimum AICc value of all the models in the set. For AIC and AICc, Δ_i represents the K-L information loss of model i relative to the best model in the set.

Burnham and Anderson (2002, p. 70–72 and 78) suggest that models with $\Delta_i < 2$ are very good models; models with $4 < \Delta_i < 7$ have less empirical support and, in most cases, models with $\Delta_i >$ about 10 can be dismissed from further consideration. Considering that the values being subtracted are often large in absolute value, it can seem odd that a difference of 10 can be so significant. Burnham and Anderson (2004, p. 271) argue that this is the case and their book presents many examples that support this conclusion, but as more experience is gained in discriminating models of complex systems, this may be reevaluated.

Posterior Model Probabilities

Posterior model probabilities are used to calculate model-averaged quantities. They are often referred to as model weights or posterior model weights. In this work they are referred to as posterior model probabilities.

Prior model probabilities are quantities defined by the user, as discussed in the section "Prior model probabilities."

In K-L information, posterior model probabilities are calculated by distributing the delta values calculated using equation 2.3 on a log probability scale. Posterior model probabilities determined in this way are also referred to as Akaike weights, and are calculated as:

$$p_i = \frac{\exp^{-0.5\Delta_i}}{\sum_{j=1}^{R} \exp^{-0.5\Delta_j}} \qquad (2.4)$$

where,

p_i is the posterior model probability, and reflects the evidence in favor of model i being the best model in the sense of minimum K-L information loss.

For five models with delta values 1, 2, 3, 5, and 10, the associated posterior model probabilities are shown in Table 1. These probabilities are used in multi-model analysis as discussed later in this chapter.

Table 1. Example delta values from Kullback-Leibler information theory and resulting posterior model probabilities, evidence ratios, and inverted evidence ratios.

Model	Delta value (eq. 2.3)	Posterior model probability (eq. 2.4)	Evidence ratio (eq. 2.5)	Inverted evidence ratio, as a percent (eq. 2.6)
1	0	0.58	1.0	100
2	2	0.21	2.7	37
3	3	0.13	4.5	22
4	5	0.078	7.4	14
5	10	0.0039	148.	0.67

Evidence Ratios and Inverted Evidence Ratios

Ratios of the posterior model probabilities for models i and j are called evidence ratios and are calculated as

$$Evidence\ ratio = p_i/p_j \qquad (2.5)$$

When i is the best model, evidence ratios can be used to make statements such as "there is 2.7 times more evidence supporting the best model (model 1) relative to the second best model (model 2)." For the example shown in Table 1, model 1 has 148 times more evidence supporting it than model 5.

While statistically correct, it can be intuitively confusing for the best model to have the smallest value. An alternative statistic is introduced in this work and is called the inverted evidence ratio. Expressed as a percent, the inverted evidence ratio is calculated as

$$Inverted\ evidence\ ratio\ as\ a\ percent = 100 \times (p_j/p_i) \qquad (2.6)$$

Inverted evidence ratios expressed as a percent can be used to make statements like "the evidence supporting model j is only 37 percent of the evidence supporting the best model (model 1)." In Table 1, the evidence supporting model 5 is less than one percent of that supporting model 1.

Evidence ratios and inverted evidence ratios are reported in the MMA main output file for all model criteria. They are always reported relative to the best model. The MMA main output file is described in chapter 6 and an example is provided in Appendix 2.

Burnham and Anderson (2002, p. 77–79) suggest that when the evidence ratio of the best and second best models is less than about 2 (equivalent to an inverted evidence ratio greater than about 50 percent), model selection uncertainty is likely to be high. That is, given other sets of data, a different model in the group is likely to be identified as "best."

Multi-Model Analysis Methods that are Consistent for k

Methods that are consistent for k approach k parameters asymptotically as the number of observations increase (Burnham and Anderson, 2002, p. 286). This suggests that the system is best described by k parameters regardless of the number of observations. These methods tend to use additional observations to more narrowly refine the selection of a given model. The consequences of this approach are investigated in the next section of this report. The BIC criterion clearly falls into this classification, and it is suspected that the same is true for KIC. This section defines these two criteria.

BIC and KIC Criteria

The Bayesian information criterion (BIC, Schwarz 1978) is calculated as

$$BIC = n\ln(\sigma^2) + k\ln(n), \tag{2.7}$$

where,

n and k are defined as for equation 2.2 (see additional comments about n in the section entitled "Including prior information on parameters" later in this chapter),

σ^2 is estimated based on maximum-likelihood theory as described by eq. 2.2b, and

$\ln$ indicates that log base e, which is called the natural log, is taken of the quantity in parentheses.

Kashyap's (1982) criterion (KIC) can be calculated as

$$KIC = (n-(k-1))\ln(\sigma^2) - (k-1)\ln(2\pi) + \ln|\underline{X}^T \omega \underline{X}|, \tag{2.8}$$

where,

ω is defined after equation 2.2c,

$|\underline{X}^T \omega \underline{X}|$ is the determinant of $\underline{X}^T \omega \underline{X}$, which equals the Fisher information matrix times s_{ML}^2 (eq. 2.2b), and

$\underline{X}$ is the sensitivity matrix and $\underline{X}^T$ is its transpose. $\underline{X}$ is sometimes called the Jacobian matrix.

$\underline{X}$ and $\underline{\omega}$ are augmented to account for prior information equations, if they are used, as discussed in the section "Including prior information on parameters" presented later in this chapter.

The use of k-1 instead of k in equation 2.8 is discussed in Appendix 4. KIC has been suggested for selection of ground-water models with increasing enthusiasm by Carrera and Neuman (1986), Neuman (2003), Neuman and Wierenga (2003), Ye and others (2004, 2005).

Additional steps

The additional steps of calculating delta values, posterior model probabilities, evidence ratios, and inverted evidence ratios, and using prior models weights are identical to the methods discussed for K-L information.

Comparison of AICc, AIC, KIC, and BIC

In practice, BIC and KIC can perform similarly to AIC and AICc; however, in some circumstances they can perform quite differently. Here we present the basic theoretical issues and then evaluate the practical implications.

A number of authors have argued that the theoretical underpinnings of BIC are philosophically weak. For example, see McQuarrie and Tsai (1998), Burnham and Anderson (2002, sections 6.3 and 6.4), and Burnham and Anderson (2004). It is argued that BIC is based on the idea that the true (or quasi-true) model exists in the set of candidate models (Burnham and Anderson, 2002 p. 284–288, 298; 2004), and the goal is to identify this model. Thus, as n increases, probability converges to 1.0 for the "true" or "quasi-true" model. BIC and other criteria that share this foundation would, then, tend to strive for consistent complexity (constant k) regardless of the number of observations. In models of natural systems (for example, ground-water models), it is argues thatBIC tends to select models that are too simple (that is, under-fitted) as the number of observations increases.

In contrast, it is argued that AICc is based on the idea that what is regarded as the best model can change as additional data are collected. The result is that more complicated models (models with more parameters) tend to be considered as additional data are collected. If it is preferable to select the model that provides the best approximation to reality for the number of observations available, the philosophical discussion suggests that this goal may be better served by the AICc criterion.

AICc and BIC both can be derived under either a Frequentist or a Bayesian framework (Burnham and Anderson, 2002, p. 284). A Frequentist views probability as the expected frequency of occurrence of an event based on observations, whereas a Bayesian views probability as a degree of belief, or in other words, a measure of the plausibility of an event given incomplete observation. Given that AICc and BIC can be derived from either viewpoint, an argument for or against a criterion should not be based on its Frequentist or Bayesian lineage. Rather, one must ask if a true (or quasi-true) model can be expected to be in the set of candidate. If so, then criteria such as BIC and KIC should be used. In cases where all models are merely approximations to complex reality, it is argued that AICc is preferable (Burnham and Anderson, 2002, p. 293). In addition, AICc is asymptotically equivalent to cross-validation (see, for example, Stone's (1977) argument for AIC), and cross-validation is a well-accepted basis of model selection (Burnham and Anderson 2002, p. 365).

KIC is similar in form to the CAICF criterion of Bozdogan (1987), which is discussed by Burnham and Anderson (2002, p. 287) and is calculated as

$$CAICF = -2\ln(\sigma^2) + k[\ln(n) + 2] + \ln |(1/\sigma^2) (\underline{X}^T \underline{\omega} \underline{X})| \qquad (2.9a)$$

Where the last term is commonly expressed as

$$\ln |(1/\sigma^2) (\underline{X}^T \underline{\omega} \underline{X})| = -(k-1)\ln(\sigma^2) + \ln |(\underline{X}^T \underline{\omega} \underline{X})| \qquad (2.9b)$$

CAICF, like BIC, is derived using the assumption that the true or quasi-true model exists in the set of models considered. In addition, CAICF is not invariant to a one-to-one parameter transformation, which is problematic in many circumstances. If the theoretical underpinnings of KIC are similar to those of CAICF, it would be expected to have some of the characteristics described above for BIC. The MLBMA method described by Neuman and Wierenga (2003) and Ye and others (2004) is achieved by using KIC as the model criterion and using equation 2.4 to calculate posterior model probabilities. Ye and others (2005) used KIC and Bayes' equation and incorporated informative prior model probability.

The practical implications of the philosophical differences discussed above can be evaluated for AIC, AICc, and BIC because of the simple form of equations 2.1, 2.2, and 2.7, respectively. KIC is can not be included in this analysis because its final term requires additional model-specific information. The evaluation proceeds noting that AIC, AICc, and BIC have the same first term. This term becomes smaller as model fit to observations improves, which generally occurs as the number of parameters increases within a similar general model concept. Then, the one or two additional terms increase for any given number of observations as the number of parameters, k, increases. Given these criteria, for a model with more parameters to be rated as preferable to a model with fewer parameters, the first term needs to decrease more than the increase in the additional one or two terms.

The number of observations, n, relative to the number of parameters, k, is important as follows. As k increases, the data are fit better (in the extreme we fit the noise in the data), but variance increases (precision decreases and confidence intervals on parameter estimates and predictions tend to be larger). This tradeoff has been discussed in a number of publications (for example, Burnham and Anderson, 2002, p. 31, 87), including in the context of ground-water models (for example, Yeh and Yoon, 1981). Resolution of the tradeoff is generally referred to as the principle of parsimony, by which a model is sought that is characterized by "…the smallest possible number of parameters for adequate representation of the data" (Box and Jenkins, 1970, p. 17; Box and others, 1994). Essentially, as parameters are added the information contained in a given set of observations is expended on more and more parameters, producing progressively less precise parameter estimates. This produces less precise predictions because the imprecision of the parameters is propagated to the predictions. An extension of the principal of parsimony to models with many parameters but in which associated smoothness criteria are imposed is discussed in the section "Multi-model averaging and highly parameterized models."

Like other model discrimination statistics, AICc can be thought of as selecting models with a balance between model fit and variance. Values of AICc calculated for a synthetic example for which a thermometer is being calibrated with "noisy" observations are illustrated in Figure 1. The models considered are (1) the mean of the measured temperatures applies for all fluid heights (model name is "mean"), (2) temperature is calculated as a linear function of fluid height (that is, $t = a_0 + a_1 \times h$, where t is temperature and h is fluid height; model name is "linear"), (3) temperature is calculated as a second-order polynomial of fluid height (that is, $t = a_0 + a_1 \times h + a_2 \times h^2$; model name is "poly2"), and (4) four other polynomial models with added terms and powers of 3, 4, 5, and 6. In this case, the linear model provides the best balance of fit and parsimony, as identified by the small AICc value. From the point of view of the physical processes involved, the linear model is also the model we would expect to be most applicable. The improved fit achieved with higher polynomial models results from the fitting of the noise in the data and does not produce a model that more effectively represents the system of concern.

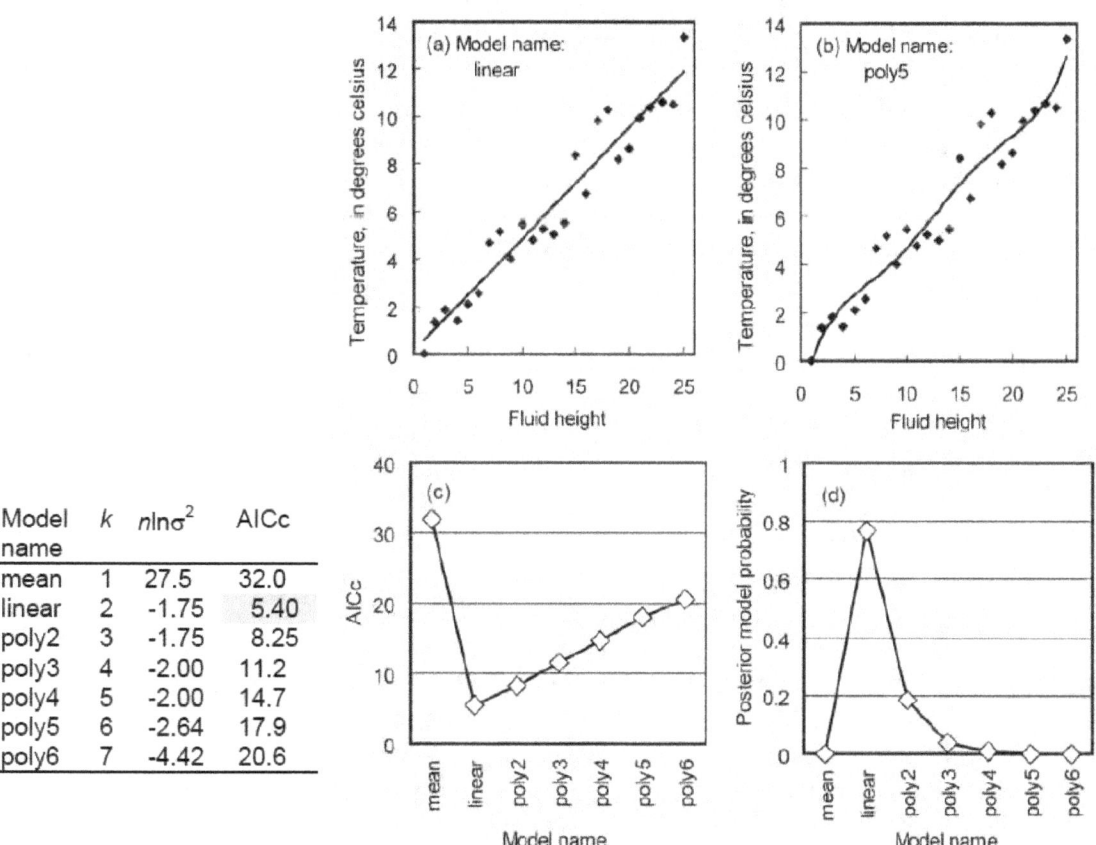

Model name	k	$n\ln\sigma^2$	AICc
mean	1	27.5	32.0
linear	2	-1.75	5.40
poly2	3	-1.75	8.25
poly3	4	-2.00	11.2
poly4	5	-2.00	14.7
poly5	6	-2.64	17.9
poly6	7	-4.42	20.6

Figure 1. An example of the how model discrimination statistics measure the balance between the better fit and reduced precision that results as parameters are added, using AICc (eq. 2.2a) as an example. The better fit is represented by the declining value of $n\ln\sigma^2$ as parameters are added (k is increased). (a) and (b) show the data and the linear and poly5 models. (c) and (d) show the AICc statistic calculated for each model and the resulting posterior model probabilities calculated using the MMA default method (equation 2.4).

The performance of the model discrimination criteria depends on the terms added to $n\ln\sigma^2$. Figure 2 shows the contributions of the additional terms to the AIC, AICc, and BIC criteria with 100 and 3,000 observations. Results for up to 98 parameters are presented in Figure 2a and 2b, which show how the criteria change as more data become available for model calibration. Results with 30 times more parameters are presented in Figure 2c, which illustrates the relative values of the criteria when the number of parameters is increased proportionately as the number of observations increases.

The results shown in Figure 2 and summarized in Table 2 demonstrate the following.
1. The third term of the AICc criterion goes to zero as the number of observations increases and the number of parameters remains constant, so that the curves for AIC and AICc overlap in figure 2b. This demonstrates how AICc approaches AIC as n/k becomes large.
2. The difference in performance of the AICc and BIC criteria as n/k increases is striking, as shown by comparing figures 2a and 2b. The following characteristics are noteworthy.
   a. As the number of observations increases, the additional terms for the AICc criterion make models with more parameters <u>much more likely</u> to be selected. For example, with 100 observations, figure 2a and Table 2 show that a model with 80 parameters would be preferred over a model with 50 parameters only if the added parameters resulted in the first term of the AICc criterion decreasing by 618 or more. That is, $n\ln(\sigma^2)|_{k=50}$ - $n\ln(\sigma^2)|_{k=80}$, where $|_{k=50}$ means evaluated at 50 parameters, would need to be 618 or more. That translates into reducing σ^2 by a factor of 484 or more. With 3,000 observations, the first term would only need to decrease by 63 or more. That translates into reducing σ^2 by a factor of 1.02 or more. Thus, given 3,000 observations, AICc is substantially more likely to be smaller for the model with more parameters than if it was given only 100 observations.
   b. As the number of observations increases, the additional term for the BIC criterion makes models with more parameters only slightly more likely to be selected. For example, with 100 observations, figure 2a shows that a model with 80 parameters would be preferred over a model with 50 parameters if the added parameters resulted in the first term of the BIC criterion decreasing by 138 or more. That translates into reducing σ^2 by a factor of 3.98. With 3,000 observations, the first term would need to decrease by 240 or more, which is a larger amount than the 138 required with 100 observations. The 240 translates into reducing σ^2 by a factor of 1.08.

3. As the number of observations and the number of parameters both increase (by a factor of 30 going from Figure 2a to 2c), the performance of the criteria again differ.
   a. The additional terms for the AICc criterion increase proportionately.
   b. The additional terms for the BIC criterion increase by more than a factor of 30, indicating that the additional observations are less likely to translate into selecting a model with more parameters when using BIC than AICc.

The examples presented in the last few paragraphs focus on the number of observations and do not address the possibility of including prior information on parameters in the criteria. For a discussion of this topic, see the text following equation 2.2 and the section entitled "Including prior information on parameters."

The results discussed under items 2 and 3 above demonstrate what was expressed theoretically in the first part of this section. As mentioned, consensus has not been reached about these criteria, and, though the authors of this work believe that AIC and AICc show more promise than BIC and KIC, the most useful approach at this time is to consider values for all four criteria and the associated multi-model inferences.

Alternative Methods

When using MMA for model selection and model averaging, methods can differ from the default procedure in four ways: (1) models can be eliminated from consideration based on an analysis of estimated parameter values; (2) the model criterion can be different; (3) prior model probabilities can be included in the analysis; and (4) posterior model probabilities can be calculated differently, which can affect the model ranks. These options are discussed in the following sections.

Eliminating Models Based on an Analysis of Estimated Parameter Values

Detecting models that are likely to produce inaccurate results using unreasonable estimated parameter values (or unreasonable relative parameter values) has been suggested by Poeter and McKenna (1995), Poeter and Hill (1996, 1997), Hill (1998), Hill and Tiedeman (2007), and others. When considering multiple alternative models, Poeter and McKenna (1995) suggest omitting such models from consideration. MMA provides support for eliminating models based on estimated parameter values through the Param_Eqns input block described in chapter 5 of this report. Models with parameter values that violate the equations defining acceptable values (or acceptable relative values) are classified as being flawed representations of the system and are omitted from further consideration.

Other Criteria Used for Model Selection

There are many other criteria for model selection (McQuarrie and Tsai, 1998). For example, Hannan and Quinn's (1979) criterion (HQ), calculated as

$$HQ = n\ln(\sigma^2) + 2k\ln[\ln(n)], \tag{2.10}$$

where, σ^2, n and k are defined as for equation 2.2a.

For HQ, sometimes the coefficient of 2 in the second term is assigned a larger value. Given that HQ is not frequently used, it is not included in the automatically calculated model measures in MMA. It can be accessed by using the CritEqn keyword in the Analyses input block described in chapter 5.

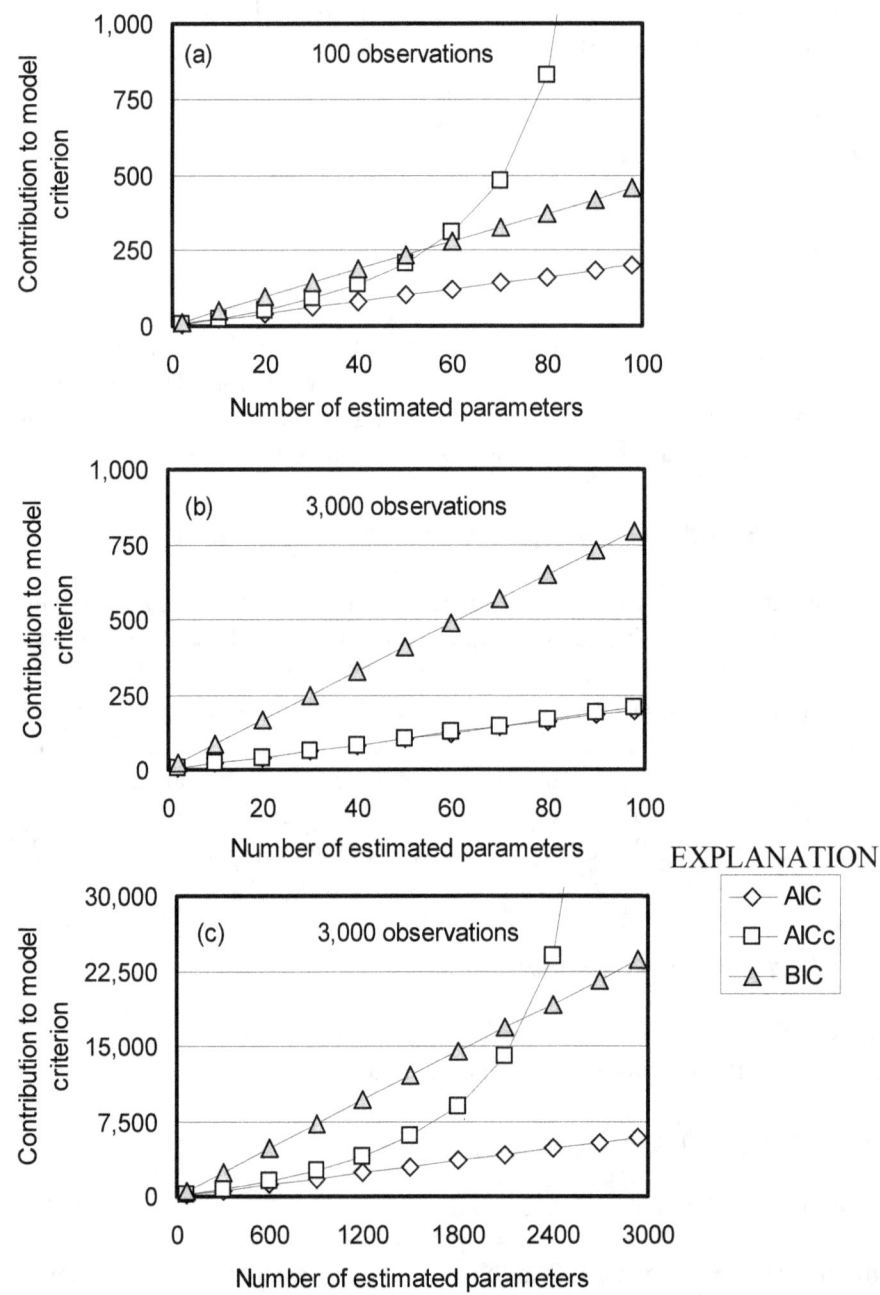

Figure 2. The contribution to the model criteria of the second and, where present, third terms in the equations for the criteria AIC, AICc and BIC (eq. 2.1, 2.2a, and 2.7). These terms increase as the number of parameters increases. (a) Results for 100 observations and up to 98 parameters. (b) Results for 30 times 100, (3,000) observations for up to 98 parameters. (c) Results for 30 times 100 (3,000) observations and up to 30 times 98 (2,940) parameters. Large values of AICc are not shown; for (a), the contribution to the AICc criterion is 10,098 for 98 parameters, and for (c), the contribution to the AICc criterion is 294,296 for 2,940 parameters]

Table 2. Required change in the term $n\ln\sigma^2$ for the criteria AICc and BIC to show preference for a 80-parameter model over a 50-parameter model.

[AICc, corrected Akaike's Information Criterion (eq. 2.2a); BIC, Bayesian Information Criterion (eq. 2.7); n, the number of observations; ln, the natural log; σ^2, the variance of a calibrated model calculated using the maximum-likelihood expression (eq. 2.2b); k, the number of parameters; $|_{k=50}$ indicates that the preceding term is evaluated at the number of parameters indicated]

| | Criterion | | | |
	AICc		BIC	
Number of observations (n)	100	3,000	100	3,000
Change in first term required to prefer 80 rather than 50 parameters ($n\ln\sigma^2|_{k=50} - n\ln\sigma^2|_{k=80}$)[1]	618	63	138	240
Ratio of variances required to achieve the value of ($n\ln\sigma^2|_{k=50} - n\ln\sigma^2|_{k=80}$)	484	1.02	3.98	1.08

[1] The fact that this decreases as the number of observations increase for AICc and increases for BIC is striking evidence of the difference between AICc and BIC.

Prior Model Probabilities

Prior model probabilities can be used to indicate, for example, that one of the hydrogeologic interpretations used to develop alternative ground-water models is considered to be more likely. The likelihood of the models needs to be based on evaluations that exclude the observations used to calibrate the model. Prior model probabilities are defined by the user in the Model_Paths input block described in chapter 5.

Under a Bayesian framework, the models each would be assigned prior model probability, τ_i; where the τ_i's for all models considered need to sum to one. If the number of models considered equals the number originally listed by the user and the user-defined prior probabilities sum to 1.0, the prior model probabilities equal the values defined by the user. If any models have been omitted from consideration for any of the reasons listed in the section "Omitting selected models from the analysis," the probabilities of the remaining models are adjusted proportionately so that the values again sum to 1.0. For example, if omitted models have prior model probabilities that sum to 0.25, the prior model probabilities of the remaining models would be multiplied by 1/(1.0-0.25) to obtain values that sum to 1.0.

If prior model probabilities are used, equation 2.3 is replaced by:

$$p_i = \frac{\exp^{-0.5\Delta_i} \tau_i}{\sum_{j=1}^{R}\left(\exp^{-0.5\Delta_j} \tau_j\right)} \tag{2.11}$$

where R is the number of models considered, prior model probabilities need to be used carefully and be supported by analysis of information not directly represented by the calibration process.

In the default option provided by MMA, prior model probabilities are not assigned. In practice, this is equivalent to setting all values of τ_i equal to 1/R.

See the section "A final note on prior model probabilities" later in this section for a discussion of how equation 2.11 relates to a true Bayesian approach.

Alternate Methods to Calculate Posterior Model Probabilities

Equation 2.4 is often used to calculate posterior model probabilities. More generally, posterior model probabilities can be calculated as:

$$p_i = \frac{PrEqn_i}{\sum_{j=1}^{R} PrEqn_j} \tag{2.12a}$$

where $PrEqn_i$ is a function of a model criterion.

If prior model probabilities are included as they are in equation 2.11, posterior model probabilities are calculated as:

$$p_i = \frac{PrEqn_i \tau_i}{\sum_{j=1}^{R} \left(PrtEqn_j \tau_j \right)}. \tag{2.12b}$$

Equations 2.12 a and b are similar to equations 2.4 and 2.11 in that they produce a normalized set of posterior model probabilities that sum to 1.0, as required.

Equations 2.12 a and b differ from equations 2.4 and 2.11 in that $PrEqn_i$ can be nearly any function of nearly any model criterion. The model criterion is defined using keyword CritEqn in the Analyses input block described in chapter 5. It is used to calculate $PrEqn_i$ as defined using keyword PrEqn in the Analyses input block. Both CritEqn and PrEqn can be defined using the MMA equation capability. In the equations for PrEqn, the expression ValCrit can be used to specify the value of the model criterion calculated for one model. Other expressions are provided so that the user can, for example, use the maximum or minimum value of the model criterion for the set of models in the equation. A complete list of expressions is provided in chapter 5.

For example, consider that the user wants to define posterior model probabilities that increase linearly from the model with the smallest value of the model criterion to the model with the largest value of the model criterion. The desired posterior model probabilities can be obtained by defining PrEqn as:

$$PrEqn_i = 1 + \left(\frac{MinCrit - ValCrit_i}{MaxCrit - MinCrit} \right) \tag{2.13}$$

where,

MaxCrit   is the largest value of the user-defined criterion for the models included,

MinCrit   is the smallest value of the user-defined criterion for the models included, and

ValCrit$_i$ is the criterion calculated for model *i*.

Posterior model probabilities are then calculated using equation 2.12 a or b, depending on whether prior model probabilities are defined.

A *Final Note on Prior Model Probabilities*

Burnham and Anderson (2002, p. 76–77) note that applying prior model probabilities as in equation 2.11 or 2.12b does not result in a true Bayesian approach unless two other steps are included. First, the prior probability distribution on the parameters needs to be included for each model. Second, derivation of posterior results requires integration that generally is achievable only by Markov Chain Monte Carlo methods, which is computationally intensive and not included in MMA version 1.000.

Prior model probability is distinct from prior parameter probability (prior parameter probability is discussed in the last section of this chapter). In a model of a ground-water system, for example, prior model probability might be determined based on an analysis of the likelihood of a proposed geologic depositional sequence for a region; or on the preference of experts for a model of recharge rate increasing as a function of elevation as compared with one generated based on slope and aspect as well. Prior parameter probability might be determined for parameters used to define hydraulic conductivity and(or) storage properties based on the results of small-scale pump tests for which the drawdown data are not included as observations when calibrating the ground-water model of concern. The relation of prior model and parameter probability has not been investigated by the statistical community to the authors' knowledge in the context of, for example the ideas presented in Burnham and Anderson (2002). This is an active area of research.

Multi-Model Inference

The traditional approach to data analysis has been to use the best model (identified based on some criteria or test result) to infer parameters estimates, predictions, and estimates of precision such as confidence intervals. Analysis has been limited to the one best model, as if no other models had been considered. Burnham and Anderson (2002, p. 150) suggest that this strategy may be adequate if one model has a posterior model probability that exceeds 0.9. Otherwise, there are multiple reasonable models, and using the one best model to calculate confidence intervals is likely to result in confidence intervals that are too narrow. In addition, confidence intervals that account for the variation in the set of models considered are likely to be more stable as new observations from the same underlying processes are introduced. This is because the various ways of representing the system are probably already included in the set of models considered, so an observation that supports one alternative over another does not drastically affect the analysis.

Chapter 2: Methods for Multi-Model Analysis

This section describes the methods MMA uses to produce model-averaged predictions and parameter values, and associated variances and linear confidence intervals.

Model-Averaged Predictions and Parameter Values

Model averaging considers predictions and optimal parameter values from multiple models. Model-averaged predictions and model-averaged optimized parameter values are calculated in a similar manner. Model averaging of predictions is discussed first because it is straightforward in that the same items are predicted using each model. In contrast, each model may not have the same parameters.

If the value of a predicted quantity differs markedly across the models (that is, the predictions differ across models with ranks $i = 1, 2, ..., R$), and no single model is clearly superior (no posterior model probability is greater than 0.90), then it is misleading to report the prediction from one model. It can be useful to list a set of predicted values or to provide a model-averaged value and statistics that reflect the range of the predictions and their uncertainty.

If model-averaged predictions are used, they are calculated as:

$$\bar{z}_q = \sum_{i=1}^{R} p_i z_{q,i} \, , \tag{2.14}$$

where,

$z_{q,i}$ is the qth predicted value for each model i,

$\bar{z}_q$ denotes the model-averaged prediction,

R   is the number of included models, and

p_i is the posterior model probability from equation 2.4, 2.11, or 2.12.

As noted in the beginning of chapter 2, item 2 of the "Overview" section, the number of models (here, R) included in the MMA analysis can be affected by a number of factors in addition to the choice of models by the modeler.

Parameter values also may vary between models, and, like predictions, may be reported as a list of values or as a model-averaged quantity with statistics that reflect the range of parameter values and their probability. If model-averaged parameter estimates are used, they can be calculated as

$$\bar{b}_j = \sum_{i=1}^{R'} p'_i \left(b_j\right)_i. \tag{2.15}$$

where,

$\left(b_j\right)_i$ is the estimate of the jth parameter produced by the ith model,

p'_i is the weight for model i calculated for the subset of R' included models, and

R'   is the subset of models in which parameter b_j is defined and estimated.

For many parameters, R' is less than R of equation 2.14 because the parameters defined in different models may represent different quantities. The results of equation 2.15 are meaningful only for parameters that represent the same entity in all R' models. See the section below entitled "Including prior information on parameters" for additional information.

MMA tests to determine whether the parameters are the same by comparing the names assigned to the parameters for each model. For example, in UCODE_2005, the names for the parameters are assigned using keyword ParamName of the Parameter_Data input block (Poeter and others, 2005, p. 69). For each model, the names are printed in a data-exchange file with filename extension _pc, and this file is read by MMA.

In nonlinear models, running the model using the model-averaged parameter values may not result in simulated values that match the model-averaged predictions. Though somewhat disconcerting, this is a direct consequence of the theory applied. Future research may develop more satisfactory approaches.

Model-Averaged Variance

The most basic expression of range is to report the largest and smallest values of predictions or parameter estimates. However, this neglects to indicate the likelihood of the models that produce the extreme values or the uncertainty of the reported values. These deficiencies can be addressed by reporting model-averaged variance or standard deviation. Variances for predictions and parameter values are presented here.

For prediction z_q, the model-averaged variance, $\overline{\text{var}}(\overline{z}_q)$, is calculated from multiple models as (Burnham and Anderson, 2002, p. 162):

$$\overline{\text{var}}(\overline{z}_q) = \left[\sum_{i=1}^{R} p_i \sqrt{\text{var}(z_{q,i} \mid \text{model}_i) + (z_{q,i} - \overline{z}_q)^2} \right]^2 \tag{2.16a}$$

where,

$$\text{var}(z_q \mid \text{model}_i) = \left(\frac{\partial z_q}{\partial \underline{b}} \right)_i^T [s^2 (\underline{X}^T \underline{\omega} \underline{X})^{-1}] \left(\frac{\partial z_q}{\partial \underline{b}} \right)_i \tag{2.16b}$$

s^2, and $\underline{\omega}$ are defined after equation 2.2 and $\underline{X}$ is defined after equation 2.8. These three quantities are calculated using model i and its optimal estimated parameter values.

For parameter b_j, the model-averaged variance is calculated from multiple models as

$$\overline{\text{var}}(\overline{b}_j) = \left[\sum_{i=1}^{R'} p'_i \sqrt{\text{var}(b_j \mid \text{model}_i) + (b_{j,i} - \overline{b}_j)^2} \right]^2 \tag{2.17a}$$

where,

$$\mathrm{var}(b_j \mid \mathrm{model}_i) = [s^2(\underline{X}^{\mathrm{T}}\,\underline{\omega}\,\underline{X})^{-1}]_b.$$ (2.17b)

The subscript b means that the variance equals the diagonal entry associated with parameter b. R' was discussed after equation 2.15.

Equations 2.16 and 2.17 include model selection uncertainty because the first term represents the variance given one model and the second term represents the variance among the set of models. It is advantageous to use the variances calculated using equations 2.16 and 2.17 even if the modeler chooses to report the prediction or parameter estimate of the "best" model rather than the model-averaged values.

The model-averaged standard deviation is calculated as the square root of the model-averaged variance.

Model-Averaged Linear Confidence and Prediction Intervals

Model-averaged linear confidence and prediction intervals can be calculated with model-averaged standard deviations using the usual procedures described, for example, by Hill and Tiedeman (2007, p. 176). For example, a linear, individual 95-percent confidence interval on a prediction identified as z_q is calculated as:

$$\overline{z}_q \pm 1.96\sqrt{\mathrm{var}(\overline{z}_q)}\,.$$ (2.18)

where the averaged terms are calculated using equations 2.14 and 2.16a.

A linear, individual 95-percent confidence interval on a parameter is calculated as:

$$\overline{b} \pm 1.96\sqrt{\mathrm{var}(\overline{b}_j)}\,.$$ (2.19)

where the averaged terms are calculated using equations 2.15 and 2.17a.

As discussed by Cooley and Naff (1990), Hill (1998), and Hill and Tiedeman (2007), the use of linear intervals needs to be accompanied by an analysis of model linearity. Such an analysis can be accomplished using the modified Beale's measure, or the nonlinearity measures developed by Cooley (2004), and discussed by Christensen and Cooley (2004), Poeter and others (2005), and Hill and Tiedeman (2007). These nonlinearity measures need to be calculated for models that dominate the calculation of the model-averaged variance; that is, for the models with the highest posterior probabilities.

If the most probable models are nonlinear, a satisfactory evaluation of uncertainty may require that nonlinear intervals be calculated using, for example, UCODE_2005, the UNC Process of MODFLOW-2000 (Christensen and Cooley, 2004), or PEST (Doherty, 2004). Nonlinear intervals are discussed by Hill and Tiedeman (2007) and references cited therein. It is not clear how to calculate model-averaged nonlinear intervals. Thus, available options include, for example, reporting nonlinear intervals for the most probable models, or reporting a combination of model-averaged linear and nonlinear intervals. Nonlinear intervals are much more computationally demanding than linear intervals.

Model-averaged and individual-model confidence intervals can be displayed as shown in Figure 3. In general, model-averaged confidence intervals are wider than individual intervals, as illustrated in figure 3. Model-averaged interval limits can be more extreme than the limits calculated for any of the individual models. Though somewhat counterintuitive, this can occur when the second term of equation 2.16a or 2.17 is large and (or) when the value produced by a model with large posterior model probability differs significantly from the values produced by other models.

Model-averaged uncertainty measures have advantages and disadvantages that in some ways are similar to those of measures produced by individual models. Basically, any errors in how the models represent the system can result in measures of uncertainty that are faulty. However, including a number of possible system representations can reduce this deficiency.

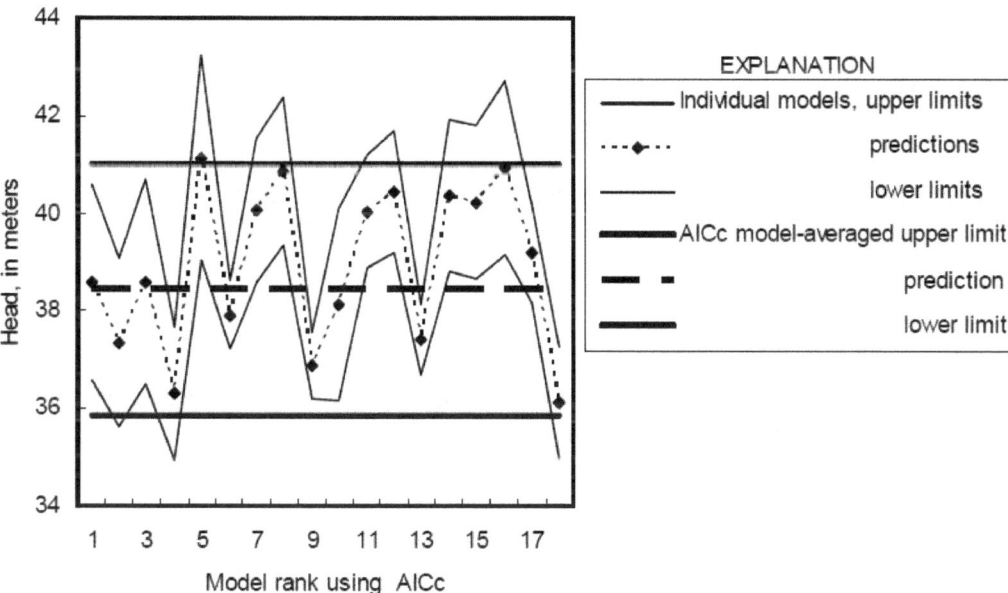

Figure 3. Predictions and interval limits for 18 models and model-averaged prediction and interval limits.

Including Prior Information on Parameters

Prior information on parameters is used to include information about parameter values developed from data other than the data reflected in the observations used in model calibration. Prior information on parameters of the type discussed here is used routinely in many fields. It has been used in ground-water modeling since the early 1980s (for example, Cooley, 1983), and is discussed in many works, including Hill and Tiedeman (2007).

MMA allows model measures to be calculated with prior information if the same prior information equations are defined for all models in the analysis. MMA determines whether the prior information equations are the same by comparing the names assigned to the prior information equations for each model. For example, in UCODE_2005, the names for the prior information equations are assigned using keyword PriorName of the Linear_Prior_Information input block (Poeter and others, 2005, p. 95). For each model, the names are printed in a data-exchange file with filename extension _pr, and these files are read by MMA. If the prior information equation names are identical for all models in the analysis, the prior information equations are used by MMA.

It is up to the user to ensure that prior information equations with the same name are defined equivalently for different models as needed. For example, prior information may be assigned for the hydraulic conductivity of all sand lenses. Models may differ in the distribution of sedimentary structures, including the sand lenses, but the prior information would be equivalent because the parameter involved consistently represents the hydraulic conductivity of the sand.

For the MMA default analyses, the following sequence of steps occurs. See chapter 6 for a description of the files mentioned.

1. It is assumed that prior information is to be included. MMA checks for the existence of prior information using data in the _dm data-exchange file produced by each model. If the number of prior information equations for each model is the same, MMA reads the _pr data-exchange file from each model.

2. If the same list of names occurs for each model, prior information on parameters is included in the calculations.

   a. The MMA output files with filename extensions ending in AICwPri, AICcwPri, BICwPri, and KICwPri are printed.

   b. The MMA output files ending with filename extensions _mma, _mma_gstats, _rank, and _rank_gstats list model measures and ranks produced using only the observations and also including prior information, and compare them. The model measures are described in chapter 3. The comparisons listed in these files support evaluation of the consistency of the observations and prior information equations. If results with observations only are very different from results using prior information on parameters, errors in the observations, prior information, or model may be indicated. The comparisons reported in these files are described in chapter 3.

3. If there is no prior information on parameters or the names of the prior information equations are not the same, the default model analyses are calculated for observations only, and MMA output files with filename extensions ending in AICObs, AICcObs, BICObs, and KICObs are printed.

If the Analyses input block is included in the main MMA input file, as described in chapter 5, the files listed in items 2a and 3 above are replaced by files with filename extensions ending with the criteria names listed in the Analyses input block. If the user selects a model measure that includes prior information and prior information equation names are the same for all models, then the analysis proceeds. If the conditions on prior information are not satisfied, a message is written and MMA continues on to the next analysis listed. The message is written to two output files, MMAroot.#mout and MMAroot._anals_AnalysisName, which are discussed in chapter 6. The files listed in item 2b above are still produced.

When prior information equations are included in the analysis, n of equations 2.1, 2.2, 2.7, and 2.8, and elsewhere in this report is defined as n=NOBS+NPR, where NOBS is the number of observations and NPR is the number of prior information equations. If the prior information equations are not used, then n=NOBS. The dimensions of the matrices $\underline{\omega}$ and $\underline{X}$ are defined using n; the matrices contain the observation weighting and the sensitivities, respectively, and are defined after equation 2.2 and after equation 2.8, respectively. Thus, the matrices are augmented when prior information equations are defined. The augmented matrices are described by Hill and Tiedeman (2007, their Appendix B).

Some Difficulties in Using Multi-Model Analysis

In practice, there are a number of complicating factors that can occur when using multi-model analysis. This section discusses three of them: (1) difficulties caused by highly parameterized models, (2) a difficulty with analyzing models with different processes that result because MMA requires the same observations and associated weighting be used for all models, and (3) inclusion of parameters for which values could not be estimated by regression.

Multi-Model Averaging and Highly Parameterized Models

In highly parameterized models, the number of parameters often exceeds the number of observations. Highly parameterized models are used in a number of fields, including geophysics (Constable and others, 1987) and ground water (Gomez-Hernandez, 2006; Franssen and others, 2003; Kitanidis, 1997; Valstar and others, 2004; Hunt and others, 2007). The parameters are generally spatially distributed within a system. When numerical methods are used to represent a system, the parameters may be located at every cell or element of a grid or mesh, or interpolation methods such as kriging may be used to obtain a complete property distribution. Of course, estimation of such a large number of parameters relative to the number of observations is problematic. A tractable problem often is obtained by requiring the parameter distribution to be as smooth as possible. This is accomplished by adding a term to the objective function shown in equation 2.2c to achieve an objective function of the following form.

$$S(\underline{b}) = \sum_{i=1}^{n} \omega_i [y_i - y'_i(\underline{b})]^2 + \lambda \sum_{\ell=1}^{nreg} \omega_{\ell+m-1} [p'_{\ell 1} - p'_{\ell 2}]^2 \qquad (2.20)$$

The variables in the first term were defined for equation 2.2c. The variables in the second term are:

λ is a factor that controls the importance of the second term,

nreg   is the number of regularization equations,

$\omega_{\ell+k-1}$ is the weight for one regularization equation (this form of the equation applies when the weight matrix is diagonal, but many programs support full weight matrices; see Hill and Tiedeman, 2007, p. 34–35 and 298 for a discussion of full weight matrices),

$p'_{\ell 1}$ is the first of the two parameter values included in the ℓth regularization equation, and

$p'_{\ell 2}$ is the second of the two parameter values included in the ℓth regularization equation, and often is within a user-defined distance of $p'_{\ell 1}$.

Including the difference between parameter values in the second term means that as the parameter values differ more from one another, the second term increases. Thus, theoretically, variation in the property distribution will occur only as needed to improve the fit as represented in the first term. The parameter pairs considered in the second term depend on the method used and often on choices made by the user.

The model fit obtained by the first term depends on the value of λ. As λ is assigned a smaller value, the penalty produced by property variability is reduced. This tends to result in greater variability in the parameter values, which in turn tends to produce a better fit to the observations and prior information represented by the y_i, and the first term of equation 2.20 becomes smaller. For example, in PEST (Doherty, 2004), the user specifies the value desired for the first term, and the value of λ is altered to achieve that value. A minimum value of λ can be defined by the user. In many situations the value of λ can be much less than 1.0.

When regularization is used, it is not clear how to use the model discrimination criteria presented in this work. If a large value of k, the number of parameters, is used without adding *nreg* to the number of observations to obtain n, large values of the model discrimination criteria are produced. If *nreg* is included in n, and λ is assigned a very small value, the *nreg* regularization equations are overemphasized in the model criteria.

One option is to use λ to adjust *nreg*, so that smaller values are added to n as λ becomes smaller. This approach is consistent with Tarantola (2005, p. 73). Further work in this area is needed to establish clear guidance for how to accommodate highly parameterized models in multi-model analysis.

Observations and Evaluation of Alternative Models with Different Processes

In some circumstances, the importance of selected dynamics to model goals may be unclear, and models with very different processes are considered. For example, in a ground-water model the following may be included in some models and not in others: transient dynamics, the effects of fluid density variations, and interaction with selected surface-water features. As noted by Hill and Tiedeman (2007, chapter 9), often different processes provide the opportunity to use different types of observations, and the need to define different types of parameters. The need for additional parameters is consistent with the theory behind MMA. If the added parameters and associated processes sufficiently improve model fit to data, that model will be preferred over models that do not include the process.

The situation is not so straightforward for the possible new observations. The theory behind MMA suggests that the models be compared using the same set of observations. The advantage of this approach is its clarity — the models are evaluated using a consistent basis. Its disadvantage is that it is not always easy to organize the data and the models such that this requirement is met. If the model averaging provided by MMA is desired, a solution needs to be found. Potential solutions depend on the situation. Here we examine two situations common in ground-water modeling and some of the alternatives that might be considered.

Stream Example

Adding a stream to a ground-water model may provide the opportunity to include more streamflow gain or loss measurements to model calibration. The first step is to investigate the models with and without the stream using the methods discussed by, for example, Hill and Tiedeman (2007, Guidelines 9 and 10), to see if adding the feature clearly improves or degrades model fit to observations and(or) estimated parameter values. If the situation is clear, the analysis with MMA can be limited to the clearly better set of models (those either with or without the feature in question). If the evaluation produces ambiguous results, analysis with MMA could be pursued for both sets independently and results compared.

To use one run of MMA to evaluate the models that both include and exclude the feature in question, a single set of observations needs to be defined. There are two options: add observations to the one set of models or omit them from the other set of models. Omitting the streamflow gain or loss observations from the one set of models is straightforward, but often is undesirable. Adding the observations to the other set of models requires devising a simulated equivalent to which the observations could be compared. If there is another mechanism by which seepage is simulated that is comparable to the stream, then the streamflow gain and loss observations might become a valuable addition to those models. If there is no seepage mechanism, a simulated equivalent of zero could be used to represent that fact that the model does not simulate streamflow. This could be achieved with UCODE_2005, for example, by creating a set of instructions that always read a zero for that observation from a file that is not changed by the model.

31

Steady-State/Transient Example

In this example both steady-state and transient models of ground-water conditions are considered. It could be that the goals of a study are well served by a model of average annual conditions, which can be represented by a steady-state model. Or, it may be argued that the goals are well served by a model that represents seasonal and(or) annual variations, which would require a transient model. Or, perhaps a sequence of steady-state models could adequately represent the system. The debate between steady-state and transient models often is valid for ground-water models because they tend to change slowly over time, so that steady-state models are sufficient in many circumstances.

Commonly, the observations for a steady-state model in such a circumstance are obtained either by (a) averaging observations at locations where there are observations over time or (b) identifying a time when observations are thought to represent average annual conditions and, as much as possible, using observations only from that time. Observation weighting ideally is determined based on an analysis of the observation errors, as discussed by Hill and Tiedeman (2007, Guideline 6). However, the transient model would use all (or many) of the observations individually, so there is a discrepancy that makes it impossible to analyze the steady-state and other models using the theory upon which MMA is based.

The basic steps described above for the streamflow example apply. That is, first check to see if one type of model is clearly superior. If that analysis does not produce a clear result, then consider evaluating the steady-state models separately from the transient models.

If an integrated analysis with MMA is desired, the observations need to be altered in some way to attain consistency. One alternative for the steady-state model is to define repeated observations at the same location of the steady-state model, one for each time there is an observation at the location in any transient models of the system. This has the disadvantage that the repeated observations at single locations might overemphasize model fit at that location in the regression, although if the transient data are quite variable, the sometimes large residuals may reduce the problem. That can be evaluated by comparing the regression results with the repeated observations with those using the set of observations for the steady-state model derived as discussed above. It might be expected that weights at each observation location could be divided by the number of observations at that location to create parity, but this method would result in the steady-state and transient models using different weights on the observations, which is not valid when using MMA.

Parameters for which the Values are not Estimated by the Regression

Commonly, the parameters represented by NPE, which is defined after equation 2.2a, are estimated by regression. However, as noted by Hill and Tiedeman (2007, p. 289), it can be advantageous to define the values of insensitive parameters for the regression, so that they do not change. Hill and Tiedeman (2007, p. 340) suggest that including such parameters in uncertainty analyses is important if the parameters are important to predictions of interest. The parameters can be included by activating them, adding prior information on them, if it is available, and rerunning the model to obtain the results required by MMA. When using UCODE_2005, the last step is accomplished with the Sensitivity-Analysis mode (see Poeter and others, 2005, p. 30).

From a pragmatic viewpoint, the procedure suggested above addresses a clear problem and appears to provide a reasonable solution. However, this situation has not been discussed in the statistics literature associated with multi-model analysis, and the procedure cannot be thought of as standard statistical practice.

Chapter 3: Variables and Measures of Model Fit Available for User-Defined Equations

This chapter presents variables and measures of model fit that are printed to MMA output files as described in chapter 6 and can be included in user-defined equations to calculate model criteria. The user-defined equations are defined as part of the Analyses input block described in chapter 5. The variables and measures of model fit are listed there for easy reference when constructing input files. This chapter provides additional explanation and background about each variable and measure of model fit printed by MMA.

Each variable and measure is assigned a label that is used as an identifier in the output files (e.g. NPE for the number of estimated parameters and NOBS for the number of observations). To include a variable or measure in an equation of the Analyses input block, use its label. The labels listed here include lower and upper case letters to make them more readable. In the MMA output file, only capital letters are used. When used in the equations of the Analyses input block, lower or upper case letters can be used.

Many of the labels listed in this chapter include a "*". The "*" is not used literally. Instead, it is replaced by characters as described in the following section. The subsequent sections of this chapter discuss the variables and measures.

Replacement of "*" with "Obs", "wPri", "_PR-O", "_%", or "_Chg" in the Labels of Variables and Measures of Model Fit

Some measures can be calculated using observations only and using both observations and prior information equations, when prior information equations are available. These measures are identified by a trailing "*" in the labels listed in the following section titles. Prior information equations are used only if they are the same for all models considered, as discussed in chapter 2 in the section entitled "Including prior information on parameters." That section also described the files produced when prior information equations are defined.

Labels with the "*" replaced by "Obs" identify measures that are calculated using only observations.

Labels with the "*" replaced by "wPri" identify measures that are calculated using observations and prior information equations.

Two other replacements for the "*" are "_PR-O" and "_%". These identify values that are designed to assist in examining the difference between the measures calculated with observations ("*" replaced by "Obs") and with observations and prior information equations ("*" replaced by "wPri"), as described in the following paragraphs. The importance of such comparisons is discussed in chapter 2 in the section entitled "Including prior information on parameters."

If prior information equations are included in the analysis, then instead of one table, four tables are printed in the data-exchange files with filename ending _mma and _mma_gstats. For the first table the "*" is replaced by "Obs", and the measures are calculated using only observations. In the next three tables the "*" is replaced with "wPri", "_PR-O", and "_%". The "wPri" table presents measures calculated with both observations and prior information equations. The "_PR-O", and "_%" tables present measures calculated as follows, with "*base*Obs" representing the value of the measure calculated using only observations, and "*base*wPri" indicating the value of the measure calculated using both observations and prior information equations. The letters "*base*" represent the base part of the label defined below. For example, *base* may be SWSR, CEV, and so on.

*base*PR-O = Change in the statistic value with prior relative to without prior

$$= (base\text{wPri-}base\text{Obs}) \tag{3.1a}$$

*base*_% = Percent change in the statistic value with prior relative to without prior

$$= 100 \times [(base\text{wPri-}base\text{Obs}) / base\text{Obs})] \tag{3.1b}$$

The values of *base*PR-O and *base*_% are positive when adding prior information equations increases the value of the measure indicated by *base*, and negative when adding prior decreases the value of the measure indicated by *base*.

In the data_exchange files produced by MMA with filename extensions _rank and _rank_gstats files produced by MMA, selected measures are used to rank the models. When prior information equations are included, three tables are printed in these files instead of one. For the first table, the "*" is replaced by "Obs", and the ranks produced using only observations are listed. In the next two tables, "*" is replaced with "wPri", "_CHG". The "wPri" table lists the ranks produced when the measures are calculated using both observations and prior information equations. The "_CHG" table indicates if the model rank changed with the inclusion of the prior information equations. The changes in rank are calculated as

$$base\_CHG = \text{Change of rank} = base\text{Obs-}base\text{wPri} \tag{3.2}$$

This yields a positive number when adding prior information makes the model more probable (the rank becomes a smaller number, such as going from 5 to 4) based on the measure involved.

See chapter 6 for additional information about the output files.

Variables

NPE – Number of Parameters Estimated

NPE is the number of parameters estimated in the regression, and a value is read for each model. As discussed after equation 2.2, k=NPE+1 is used to calculate some model criteria to account for estimating the true error variance as a parameter.

Additional considerations are discussed in the section of chapter 2 entitled "Parameters for which the values are set for the regression."

NOBS – Number of Observations

NOBS is the number of observations used in the regression, and equals the variable n used in the equations of chapter 2 when prior information is not considered. A value is read for each model and only models with the largest value are used by MMA. See the following paragraph for additional discussion of why some models might have fewer observations. For the models considered, the names assigned to observations are read and need to be the same for MMA to proceed. This check helps to ensure that analyses are conducted only if the same observations and associated weighting are used for all models.

Sometimes the simulated equivalent to an observation can not be calculated so that the observation does not appear in the results of one or more models, but does appear in the results of other models. For example, in a ground-water model, variations of a water table represented as a free surface can result in the inability to calculate hydraulic head in some parts of the grid. In those areas, it is not possible to obtain simulated equivalents to head observations. MMA only can consider models with the same set of observations, so models for which one or more observation(s) do not appear are omitted from consideration and a message is printed to the main output file. If this occurs, MMA omits the model from consideration and there are three possible ways to proceed:

(1) Results produced with the reduced set of models can be used.

(2) The occasionally omitted observations can be omitted from all models. This option can be useful if the omitted observations are not among those that are most important to the regression or to the predictions, as determined using local or global sensitivity-analysis methods. Global methods are discussed, for example, by Saltelli and others (2000, 2004). Local methods are discussed, for example, by Hill and Tiedeman (2007). If the observations are not important, the regressions generally do not need to be repeated with the reduced set of observations, but, if they are not repeated, a sensitivity analysis run is required to produce files with the reduced number of observations as needed by MMA.

(3) The models with the omitted observations can be altered so that the observations are retained. For example, in the ground-water situation discussed above, making model layers thicker can alleviate the problem. In some cases a substitute simulated equivalent may be used, such as the head from an underlying grid cell that is not dry, or the elevation of the bottom of the dry aquifer. Alternatively, the free surface often can be adequately approximated in a way that reduces or eliminates the numerical difficulties of concern (Hill, 2006). Of course, the change in the model would require that the regression be repeated.

NPR – Number of Prior Information Equations

NPR is the number of prior information equations on the estimated parameter values, and a value is read for each model. Prior information equations can be different for different models analyzed using MMA, but need to be the same for the fit to prior information to be included in the analysis. Two tests are conducted to determine if the prior information is the same for all models. First NPR needs to be the same nonzero value for all models. Second, the names of the prior information equations defined for each model need to be the same. For more information, see the section in chapter 2 entitled "Including prior information on parameters."

Measures of Model Fit

The measures of model fit are divided into three categories: overall measures of model fit, statistics from graphs that compare two quantities, and statistics for evaluating bias in time and space. A "*" after the measure indicates that it can be calculated without and with consideration of prior information, as discussed in the beginning of chapter 3 in the section "Replacement of "*" with "Obs", "wPri", "_PR-O", "_%", or "_Chg" in the Labels of Variables and Measures of Model Fit."

Overall Measures of Model Fit

SWSR* – *Sum of Weighted Squared Residuals*

SWSR equals the sum of squared weighted residuals. It is equivalent to $S(\underline{b})$, which was defined in equation 2.2c.

This statistic can be calculated in two ways. First, it can be calculated using observations only, for which the "*" is replaced to produce the statistic label "SWSRObs". Alternatively, it can be calculated using observations and prior information, for which the "*" is replaced to produce the statistic label "SWSRwPri".

CEV* – *Calculated Error Variance*

CEV equals the regression estimate of σ^2 and is referred to as the calculated error variance by, for example, Hill and Tiedeman (2007). It differs from the maximum-likelihood estimate of equation 2.2b in that it has been corrected for bias that occurs when k/n is relatively large. When only observations are used, CEVObs is calculated as

$$\text{CEVObs} = \frac{SWSRObs}{(NOBS - NPE)} \tag{3.3}$$

where $SWSRObs$ equals the SWSR statistic described above evaluated using only observations, which is given the label SWSRObs.

When observations and prior information are used, CEVwPri is calculated as

$$\text{CEVwPri} = \frac{\text{SWSRwPri}}{((NOBS + NPR) - NPE)} \tag{3.4}$$

where SWSRwPri equals the SWSR statistic described above evaluated using observations and prior information, which is represented using the statistic label SWSRwPri.

A difficulty arises when models are highly parameterized because it is not clear what role *nreg* of equation 2.20 should play in equation 3.4. For additional discussion, see the section "Multi-model averaging and highly parameterized models" of chapter 2.

MLOF* – *Maximum Likelihood Objective Function*

MLOF equals the maximum likelihood objective function as defined by Burnham and Anderson (2002, p. 12), among others. It is the first term in equations 2.1, 2.2a, and 2.7, where it is expressed as "$n \ln(\sigma^2)$". A modified version of this term also comprises the first term in equation 2.8, which is "$(n-(k-1)) \ln(\sigma^2)$." Using the labels defined in the current chapter, the maximum likelihood objective function is calculated as the following when only observations are included

$$MLOFObs = (NOBS) \ln(SWSRObs/NOBS) \tag{3.5}$$

where SWSRObs equals the SWSR statistic described above evaluated using only observations SWSRObs. When there are also prior information equations,

$$MLOFwPri = (NOBS + NPR) \ln (SWSRwPri/(NOBS+NPR)) \tag{3.6}$$

where SWSRwPri equals the SWSR statistic described above evaluated using observations and prior information, SWSRwPri.

In maximum likelihood theory, SWSRObs/*NOBS* or, if there are prior information equations, SWSRwPri/(*NOBS* + *NPR*) are used to estimate the calculated error variance, as presented in equation 2.2b. However, these estimates are biased when *NPE/NOBS* or *NPE/(NOBS+NPR)* , respectively, are large, and CEV* is used in many circumstances.

AIC* – *Aikaike's Information Criterion*

AIC was presented in equation 2.1.

This measure can be calculated in two ways. First, it can be calculated with observations only using equation 3.5, in which case the label is "AICObs". Alternatively, it can be calculated with observations and prior information using equation 3.6, in which case the label is "AICwPri".

AICc* – *Akaike's Modified Information Criterion*

AICc was presented as equation 2.2.

This measure can be calculated in two ways. First, it can be calculated with observations only using equation 3.5, in which case the label is "AICcObs". Alternatively, it can be calculated with observations and prior information using equation 3.6, for which the label is "AICcwPri".

BIC* – *Bayesian Information Criterion*

BIC was presented as equation 2.7.

This statistic can be calculated in two ways. First, it can be calculated using observations only with equation 3.5, in which case the label is "BICObs". Alternatively, it can be calculated with observations and prior information using equation 3.6, for which the label is "BICwPri".

KIC* – *Kashyap's Criterion*

KIC was presented as equation 2.8.

This statistic can be calculated in two ways. First, it can be calculated with observations only using SWSRObs, in which case the label is "KICObs". Alternatively, it can be calculated with observations and prior information using SWSRwPri, for which the label is "KICwPri".

XTwX* – *Natural log of the Determinant of $\underline{X}^T \underline{\omega}\, \underline{X}$*

XTwX equals the natural log of the determinant of the NPExNPE matrix that results from the matrix product $\underline{X}^T \underline{\omega}\, \underline{X}$. When only observations are used, the statistic is calculated as

$$\text{XTwXObs} = \ln |\underline{X}^T \underline{\omega}\, \underline{X}| \tag{3.7}$$

When prior information is included, the measure is calculated as

$$\text{XTwXwPri} = \ln |\underline{X}_{aug}^T \underline{\omega}_{aug}\, \underline{X}_{aug}|. \tag{3.8}$$

In equation 3.8, $\underline{X}_{aug}$ and $\underline{\omega}_{aug}$ are augmented with prior information as shown by Hill and Tiedeman (2007, their Appendix B).

Statistics from Graphs that Compare Two Quantities (The statistic labels begin with R2, Slp, or Int)

Model fit to observations can be evaluated graphically as discussed in many works, including Cooley and Naff (1990), Draper and Smith (1998), Helsel and Hirsch (2002), and Hill and Tiedeman (2007). MMA allows statistics related to six types of two-quantity graphs to be used in the equations of the Analyses input block described in chapter 5. Four of the graphs are discussed in this section. The other two involve plotting residuals or weighted residuals against time, and are discussed in the following section of this report.

The labels for the statistics associated with the graphs begin with one of three prefixes that identify the correlation (R2), slope (Slp) or intercept (Int) of a straight line fit through the points on the graph. The statistics are labeled using R2, Slp, or Int followed by an underscore and letters that identify the graph. For three graphs, there are statistic labels with all three of the prefixes R2, Slp, and Int. For the fourth and fifth graphs, only the correlation (R2) is calculated.

Each statistic label ends with an underscore and letters that identify the graph. The data for all of the graphs come from data-exchange files such as those created by UCODE_2005. The data-exchange filenames end with a filename extension that identifies the data involved. Extensions of the relevant files include: _os, _ws, _ww, and _nm. The statistic labels use the same extension as the data-exchange files that contain the pertinent data. The file extensions used in this way are _os, _ws, _ww, and _nm. All of the graphs can be constructed using prior information so, as presented below, these endings appear with a trailing "*", indicating "Obs" or "wPri" as discussed in the beginning of this chapter. The data related to prior information are listed in the last lines of the four data-exchange files mentioned above. Additional attributes of these graphs are discussed by Hill and Tiedeman (2007, chapter 6).

For most of the graphs, MMA calculates the value of the statistics using the data from the data-exchange files mentioned above. However, this is not true for the _nm data-exchange file, as discussed below.

Example graphs are shown in figure 4. Correlation coefficients (statistic prefix R2), slopes (Slp), and intercepts (Int) for figure 4a, b, and c are calculated by MMA. For figure 4d, only R2 is reported. If prior information were included in the analysis, points associated with the prior information would be plotted as well.

Statistic Labels Ending with _os — Simulated Equivalents and Observed Values*

An example graph of these quantities is shown in Figure 4a. The data are obtained from the data-exchange file with suffix _os.

It is desirable to achieve a one-to-one relation between unweighted simulated equivalents and unweighted observed values with a correlation (R2_os*) of 1.0, slope (Slp_os*) of 1.0, and intercept (Int_os*) of 0.0. The "*" is replaced by "Obs" or "wPri", as described at the beginning of this chapter.

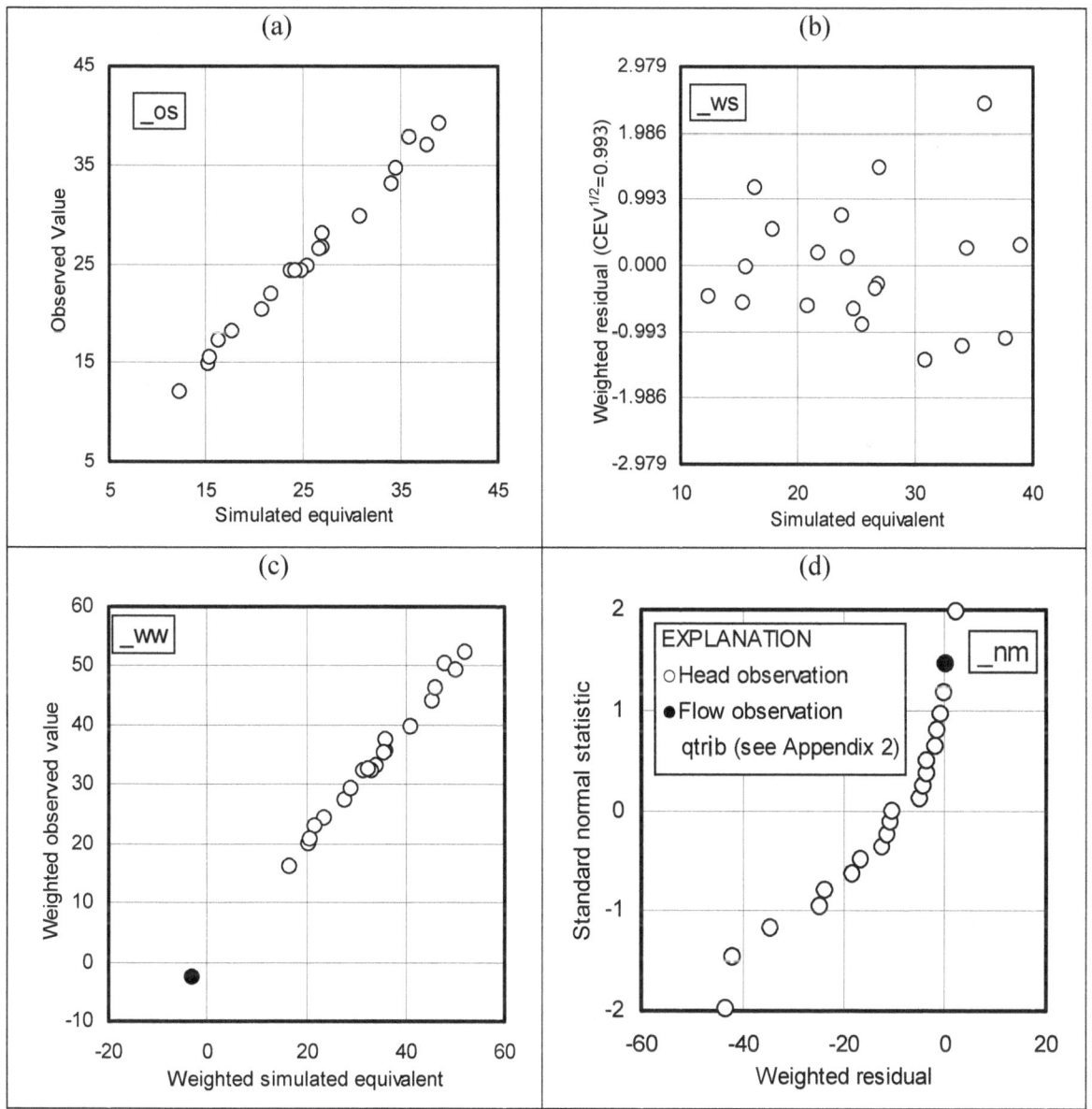

Figure 4. Graph of (a) observed values and simulated equivalents (statistics describing this graph are identified by extension _os*), (b) weighted residuals and simulated equivalents (identified by extension _ws*; the calculated standard error of the regression is used to label the vertical axis, as suggested by Hill and Tiedeman, 2007, p. 102–103, figures 6.1 and 6.2), (c) weighted observed values and weighted simulated equivalents (identified by extension _ww*), and (d) a normal probability plot, for which the statistics R2_NM* are derived. An explanation of the "*" is provided in the text. The values plotted are from model 2A of the set of models discussed in Appendix 2. Observation qtrib is not plotted in (a) or (b) because, with a simulated equivalent of –5668 and an observed value of –6992, unweighted values can not be plotted readily.

Statistic Labels Ending with _ws — Simulated Equivalents and Weighted Residuals*

Weighted residuals are calculated as

$$\omega_i^{\frac{1}{2}} (y_i - y_i{}'(b)) \tag{3.9}$$

when the weight matrix is diagonal. For a full weight matrix,

$$\underline{\omega}^{\frac{1}{2}} (\underline{y} - \underline{y}{}'(b)). \tag{3.10}$$

Variables ω_i, $\underline{\omega}$, y_i and $y_i{}'(b)$ are defined after equation 2.2c, and $\underline{y}$ and $\underline{y}{}'(b)$ are vectors containing all the observations and simulated equivalents, respectively. If prior information is included, the subscript i of equation 3.9 would include reference to the prior information equations, and the weight matrix and the vectors of equation 3.10 would be augmented for prior information.

The simulated equivalents are represented using the symbol

$$y_i{}'(b) \tag{3.11}$$

Unless weights are calculated using coefficients of variation, weighted residuals are theoretically independent of weighted residual (Hill and Tiedeman, 2007, p. 100–104).

An example graph of these quantities is shown in Figure 4b. The values are obtained from the data-exchange file with filename extension _ws.

It is desirable to find optimal parameter values that yield a narrow band of randomly distributed weighted residuals centered on zero, with uniform width for all simulated equivalents. Consequently, a correlation (R2_ws*) near 0.0, a slope (Slp_ws*) near 0.0, an intercept (Int_ws*) near 0.0, and are desirable. The "*" is replaced by "Obs" or "wPri," as described at the beginning of this chapter.

Statistic Labels Ending with _ww — Weighted Simulated Equivalents and Weighted Observed Values*

An example graph of these quantities is shown in Figure 4c. The values are obtained from the data-exchange file with filename extension _ww.

It is desirable to achieve a one-to-one relation between weighted simulated equivalents and weighted observed values with a correlation (R2_ww*) of 1.0, a slope (Slp_ww*) of 1.0, and an intercept (Int_ww*) of 0.0. The "*" is replaced by "Obs" or "wPri", as described at the beginning of this chapter.

Statistic Labels Ending with _nm — Graphical Test for Independence and Normality of Weighted Residuals*

An example normal probability graph is shown in Figure 4d, and is used to evaluate whether the weighted residuals are independent and normally distributed. The data of concern for these statistics are listed in the data-exchange file with extension _nm. The statistics are read from the data-exchange file with filename extension _dm.

For least-squares regression, it is desirable to find optimal parameter values that yield normally distributed weighted residuals (Hill, 1998, p. 23; Hill and Tiedeman, 2007, p. 108). Although typically the regression produces correlated weighted residuals, as discussed by Draper and Smith (1998), Cooley and Naff (1990), Hill (1992, 1994, 1998, p. 23–24) and Hill and Tiedeman (2007), checking for independent, normally distributed weighted residuals is easy; and if the residuals satisfy this more rigorous test, it is rarely important to consider additional evaluations. The independence and normality of the weighted residuals are evaluated by calculating correlation coefficients between weighted residuals and standard normal deviates. This correlation coefficient is represented by R_N^2 or RN2 in Hill (1998), Hill and Tiedeman (2007), and the UCODE_2005 documentation.

A correlation, R2_nm* near 1.0 is desirable. The "*" is replaced by "Obs" or "wPri", as described at the beginning of this chapter.

Statistics for Evaluating Bias in Time and Space

Residual bias in time or space is important to modeling of earth systems, but is not taken into account by the commonly used model selection criteria discussed in chapter 2. A number of measures related to temporal and spatial bias are included in MMA to help users develop model rankings that consider residual distribution in time and space.

The next section describes the measures related to time; the following section describes the measures related to space. These measures are calculated only if the MMAroot.xyzt file exists in the directory where MMA is executed. The construction of the MMAroot.xyzt file is described in chapter 5.

Time (The statistic labels begin with R2, Slp, or Int)

Like the overall measures of model fit described above, the measures designed to detect bias in time are related to graphs of data. Here, the graphs of concern are graphs of residuals and weighted residuals in relation to time (identified by statistic labels that end with _rt or _wrt, respectively), and the statistics are the correlation, slope, and intercept of those graphs (identified by statistic labels that begin with R2, Slp, or Int). Times are obtained from the MMAroot.xyzt file described in chapter 5. If all the times have the same value, then these statistics are not calculated.

Statistic Labels Ending with _rt — Residuals and Time

The residuals are obtained by MMA from the data-exchange file with extension _r.

For transient models, it is desirable to find optimal parameter values that yield a narrow band of residuals centered on zero where the width of the band does not change with time. The width of the band is expected to be uniform if the observation error has the same variance for all times. No statistic is calculated to detect variations in width with time, so the graph would need to be inspected visually to find such variations. A correlation (R2_rt) near 0.0, a slope (Slp_rt) near 0.0, an intercept (Int_rt) near 0.0 are desirable.

Statistic Labels Ending with _wrt — Weighted Residuals and Time

The weighted residuals are obtained by MMA from the data-exchange file with extension _w. The observation times are obtained from the xyzt file described in chapter 5.

For transient models, it is desirable to find optimal parameter values that yield a narrow band of weighted residuals centered on zero, with uniform width for all times. A correlation (R2_wrt) near 0.0, a slope (Slp_wrt) near 0.0, and an intercept (Int_wrt) near 0.0 are desirable. No statistic is calculated to detect variations in width with time, so the graph would need to be inspected visually to find such variations.

Space

If the xyzt file described in chapter 5 is available, the listed x, y, and z coordinates for observations are used by MMA to calculate the following basic set of statistics that indicate randomness of the weighted residuals in space. This set of statistics is not exhaustive and may not be useful in some situations. They are presented only as a first step toward addressing what can be a very difficult problem.

In the following definitions, a lower case i is used to indicate that the statistic can be calculated using the x, y, or z coordinates.

The advantage of these statistics is that they are simple and provide a quantitative summary of the visual input the user obtains by viewing the distribution of residuals in space and time. The disadvantage is that a model that has a concentric bias may not be detected by these measures.

CNT_LOCi

The centroids of the x, y, or z coordinates of the observations are calculated by summing the coordinates in each direction and dividing by the number of observations. The following provides an example.

	x	y	z
	5	4	5
	2	3	10
	11	2	3
CNT_LOCi	6	3	6

DIFCNT_Wi and DISTCNTW

Unevenly distributed weighted residuals can be detected by summing the products of the weighted residuals and their coordinates in each direction and dividing by the number of observations. The resulting statistic is referred to as CNT_Wi, and DIFCNT_Wi = CNT_Wi - CNT_LOCi. Unevenly distributed weighted residuals in the i direction are indicated by values of DIFCNT_Wi that are large in absolute value. A single summary statistic can be calculated by summing over the three coordinate directions to obtain:

$$\text{DISTCNTW} = \left(\sum_{i=1}^{3} \left(DIFCNT\_Wi \right)^2 \right)^{1/2} \tag{3.12}$$

DIFCNT_SGNWi and DISTCNT_SGNW

These are similar to DIFCNT_Wi and DISTCNTW, but account only for whether the weighted residual is negative or positive. Any negative weighted residual is replaced by –1.0; and positive weighted residual is replaced by +1.0. Values equal to zero are not changed.

DIFCNT_MGWi and DISTCNTMGW

These are similar to DIFCNT_Wi and DISTCNTW, but account only for the magnitude of the weighted residual. The absolute values of the weighted residuals are used in the calculations.

Chapter 4: MMA Execution

This chapter discusses the command needed to run MMA, provides some context and general comments about input and output files, and describes methods for troubleshooting common problems.

Running MMA

The Run Command for MMA needs to be executed from the directory containing the MMA input files. The run command form is:

> *path:*\mma *inputfile_name MMAroot*

> where:

> *path:*\ = the relative or absolute path to the MMA.exe on your computer (alternatively, this could be specified in the computer registry)

> *inputfile_name* = the name of the main input file (chosen by the user)

> *MMAroot* = the root name of the mma output files and one optional input file (chosen by the user).

Input Files

MMA requires one main input file; this is the only file that needs to be constructed by the user. Input instructions are provided in chapter 5 of this report. To conduct the default set of four analyses, this file only needs to contain the Model_Paths input block.

An optional file, MMAroot.xyzt, is needed to calculate the statistics described for characterizing the bias of residuals and weighted residuals in time and space in the section in chapter 3 entitled "Statistics for evaluating bias in time and space." These evaluations are conducted if the xyzt file is located in the directory from which MMA is executed. The xyzt file is discussed further at the end of chapter 5.

MMA also uses numerous data-exchange files from each model being considered. The data-exchange files for each model need to be located in the directory specified using the PathAndRoot keyword of the Model_Paths input block discussed in chapter 5. Commonly, results from each model are in a separate directory.

The data-exchange files needed by MMA are listed in Table 3. Contents of the data-exchange files are defined by Poeter and others (2005) and Banta and others (2006) and described briefly in Table 3. The data-exchange files are produced by UCODE_2005 when keyword DataExchange=yes in the UCODE_Control_Data input block and either sensitivities are calculated or a regression is executed (that is, UCODE_2005 modes Sensitivity-Analysis or Parameter-Estimation, respectively; see Table 3 of Poeter and others, 2005, p. 30).

Table 3. Data-exchange files produced for each model and items read by MMA.

Filename extension[1]	Contents
Always used by MMA	
_dm	Information related to model structure, fit and parsimony, including the model name; the units used for mass, length, and time; the number of defined and estimated parameters; the number of observations and prior information equations; the values of a variety of model criteria, among other variables.
_pc	Optimized parameter information, including parameter name, estimated value, standard deviation, and linear individual confidence interval.
_ss	Sum of weighted squared residuals, calculated for just observations, just prior information, and the total.

The following three files each have four columns. The first two columns contain the listed data. Column 3 contains integers that can be used to assign different plot symbols to the data when plotted. Column 4 contains names of the observation or prior information.

_os	Unweighted simulated equivalents are in column 1; observations or prior information values are in column 2.
_ws	Simulated equivalents are in column 1; weighted residuals are in column 2.
_ww	Weighted simulated equivalents are in column 1; weighted observations and prior information values are in column 2.

The following two files each have three columns. The first column contains the listed data. Column 2 contains integers that can be used to assign different plot symbols to the data when plotted. Column 3 contains names of the observation or prior information.

_r	Unweighted residuals
_w	Weighted residuals

Used by MMA when prior information on parameters is defined

_pr	Names of prior information equations.

Used by MMA when predictions are analyzed

_linp	Predictions and their linear confidence intervals.

[1] Filepaths are constructed starting with the paths and roots defined using the PathAndRoot keyword of the Model_Paths input block discussed in chapter 5, a period, and the filename extensions listed here. So, for example, for PathAndRoot= C:\model\m1\dv1, data-exchange files would be named C:\model\m1\dv1._dm, C:\model\m1\dv1._pc, and so on.

Output Files

Numerous files may be output by MMA. The user can elect to print some of these files using the Output_Control input block, or by requesting options such as parameter or prediction averaging. Output files begin with the root name specified on the program command line for MMA and end with a period followed by an MMA-defined extension. Output files are described in chapter 6.

Troubleshooting

As shown in chapter 5, the input generally needed by MMA is not very complicated. Most problems encountered in using MMA arise from typographical errors in the input file, providing a path/root that does not exist, or not providing the required files from the regressions.

Some models listed may not be included in the analysis. See the section in chapter 2 entitled "Omitting selected models from the analysis" for more information.

When model averaging is requested for a group of models for a parameter name, the parameter needs to have been estimated for each model and have the same parameter name in each model. For results to be meaningful, the parameter needs to represent the same entity in each model.

When model averaging is requested for predictions, the predictions need to have been included for every model in the evaluation. The names of the predictions need to be the same for all evaluated models.

MMA has been programmed to recognize certain errors caused by problems related to input data for the model evaluation, and problems related to the files available from the regressions. The input instructions describe some of these errors. When a problem is encountered, an error message is written to the main MMA output file (MMAroot.#mout, where MMAroot is replaced by the root defined by the user on the MMA command line). Execution is either modified or stopped.

MMA also detects situations that could indicate a problem, but may not. For example, if prior model probabilities are defined but do not sum to 1.00, the following message is printed:

WARNING: SUM OF PRIOR MODEL PROBABILITIES IS NOT 1.00

SUM = value

FOR MODELS WITH NO USER-ASSIGNED PROBABILITY, THE PROBABILITY
WAS SET TO 1/(NUMBER OF MODELS) AND USED IN THE SUM

IF THE SUM OF PRIOR MODEL PROBABILITIES IS BETWEEN
0.999 and 1.001 THE PROBABILITIES ARE NORMALIZED.

This warning message is typical of the amount of information provided to the user about potential difficulties identified by MMA.

When problems arise, it is useful to peruse the MMAroot.#mout file for information that may lead to resolution. Increasing the value of VERBOSE in the Options input block will increase the amount of material written to MMAroot.#mout and may facilitate diagnosing an input problem.

Model evaluations and multi-model inference (model averaging) are only meaningful if all of the regressions are performed in the same system of units. The fn._dm files are checked to confirm this. Evaluation is terminated if the units do not match. This is only as reliable as the entry of unit types by users performing the regressions.

Chapter 5: Instructions for the MMA Main Input File and the MMAroot.xyzt File

As described in chapter 4, MMA requires (1) a main MMA input file for which the path is defined in the MMA run command and (2) data-exchange files produced for each model. Optionally, the user can include a file describing the space/time coordinates of the observations in the same directory as the main MMA input file. Construction of this optional file is described at the end of this chapter.

Files needed from each model directory are data-exchange files generated by the regressions for which results are being analyzed by MMA, as discussed in chapter 4 and listed in Table 3.

Identifying the MMA Main Input File and MMAroot

As discussed in chapter 4, the name of the main input file for MMA and the specification for MMAroot are identified on the command line for MMA.

The MMA main input file is constructed using the input blocks described in this chapter.

The next section describes general characteristics of the input blocks and defines the keywords used in the MMA input blocks.

MMA Main Input File—A File Composed of Input Blocks

The MMA main input file is composed of up to eight input blocks. To accomplish the four default analyses described in chapter 2, only the Model_Paths input block is needed.

Each input block is composed of keywords and data applicable to the situation being considered. All input is case-insensitive. General features of the input blocks are described in the following sections, subsequent sections define the keywords available for use in each of the eight MMA input blocks. Finally, protocols for defining the equations that can be specified in some of the input blocks are presented.

The input blocks described in this report are designed using the conventions established by the JUPITER API (Banta and others, 2006), and were programmed using the API modules. See Appendix 1 for additional information about the JUPITER API.

Basic Structure of Input Blocks

The main input file includes input blocks with the basic structure shown in Figure 5. BEGIN and END need to appear literally, though lowercase letters can be used because all input items are case-insensitive. In addition, all input items are space-delimited. Square brackets are used to identify optional variables. Blocklabel, blockformat, and blockbody are defined in the following sections.

```
BEGIN blocklabel [blockformat]
      Blockbody
END blocklabel
```

Figure 5. Basic structure for input blocks. *BEGIN* and *END* need to appear literally. *blocklabel* is replaced by the name of one of the eight MMA block labels. *blockformat* is replaced by Keywords, Table, or Files, which are described below. *Blockbody* contains data; the data included, and their organization, are defined based on the alternatives chosen for *blocklabel* and *blockformat*.

Blocklabel

The variable *blocklabel* shown in Figure 5 needs to be replaced by one of the eight alternatives listed in Table 4. The *blocklabel* alternatives are described fully later in this chapter; this section provides general information

If a *blocklabel* is misspelled, the data are ignored and defaults are assigned. Ignoring unrecognized input blocks allows different applications of the JUPITER API to use the same or very similar input files. The drawback is that an input block is ignored if the *blocklabel* is misspelled. To check, review the echo of the input printed in the MMA main output file.

Of the *blocklabel* alternatives listed in Table 4, only input block Model_Paths is always needed; it identifies the models to be evaluated. If other input blocks are omitted, defaults are used. Additional capabilities of MMA are accessed by using the additional input blocks. The order in which the input blocks appear in the input file is important; Table 4 defines the required order.

Table 4. Blocklabel alternatives of the MMA main input file.

[Bold type and grey shading identify required input blocks]

Order[1]	Purpose	blocklabel	Default column order[2]
1	Define level of debugging output printed	Options	No
2	Control files to be printed	Output_Control	No
3	Define groups of models and indicate whether the following apply for the group: (1) parameter averaging and (2) omitting models for which parameter values are unreasonable, as defined in the Param_Eqns input block	Model_Groups	No
4	Define expressions for unacceptable parameter values	Param_Eqns	Yes
5	Identify parameters to be model-averaged and associated models	Param_Avgs	Yes
6	List models to be used in the analysis, defined by using the paths to models including the root of the filename.	**Model_Paths**	Yes
7	Define predictions to be model-averaged	Preds	Yes
8	Define analyses for which model ranking, weighting, and averaging will be undertaken	Analyses	Yes

[1] Order needed in the main input file.

[2] "Yes": the input block has a default column order. With blockformat TABLE, these blocks can contain data without column labels for selected keywords if the data are in default order. "No": the input block does not have a default column order. In the latter case, column labels always are needed when blockformat TABLE is used.

Blockformat

The variable **blockformat** shown in Figure 5 needs to be replaced by one of the three alternatives listed in Table 5. If it is omitted, **blockformat** KEYWORDS is used. Although a default is available, it is urged that the **blockformat** be designated explicitly to reduce confusion. The options available for **blockformat** are discussed further in the following section **Blockbody**.

If the **blockformat** specified does not match the format of the data, the information in the data block is ignored and generally no error message is printed. For example, if **blockformat** 'KEYWORDS' is specified by default or designation, data organized in 'TABLE' format is

51

ignored. The problem can be detected by inspecting the echo of the input in the main MMA output file. This is a consequence of the flexibility of the input block structure described above in the section "Blocklabel."

Table 5. Blockformat alternatives.

Blockformat[1]	Prescribed input format
KEYWORDS	Blockbody consists of a series of lines of the form: Keyword=data Under some circumstances there are restrictions on how the keywords are ordered. See the text for additional information. If no blockformat is specified, KEYWORDS is used. Comments are allowed.[2]
TABLE	Blockbody consists of a table of data that may have labels on the columns and may be read from the main input file or from another input file. See the text for additional information. Comments are not allowed.
FILES	Blockbody consists of the pathname for one or more files. Comments are allowed.[1,2] To allow the format to be specified, the contents of each of the listed files needs to begin with a 'BEGIN Blocklabel [Blockformat]' line and end with an 'END Blocklabel' line. The Blocklabel needs to be the same as in the 'BEGIN Blocklabel FILES' block within which the files are listed. This option is rarely used for MMA. For an illustration of this, see the example provided for UCODE_2005 in Poeter and others (2005, the section entitled "Observation_Data Input Block."

[1] The options described in the table will suit most users. Additional flexibility is described in the JUPITER API documentation (Banta and others, 2006).
[2] Comments can be inserted as separate lines starting with a # in the first column.

Blockbody

The **Blockbody** of Figure 5 contains data or the names of files from which the data are to be read. The format of the data is determined by **blockformat**.

The meaning of the data provided is defined using keywords. Keywords that are not recognized are ignored. This allows a constructed input block to be used for multiple purposes without modification. It also means that misspelled keywords are not flagged as errors and default values will be used if keywords are misspelled. This problem can be identified by reviewing the echo of the input file in the main MMA output file produced when keyword Verbose of the Options input block is set to 3, 4, or 5. For many keywords, a default value is available and is used if the keyword is omitted.

Some keywords can appear in any order while other keywords indicate the need for associated data to be provided either through a subsequent set of keywords or by other means. The options available depend on the blocklabel.

An example of a keyword that indicates the need for associated data occurs in MMA for blocklabel Model_Paths. Each time the keyword PathAndRoot appears, the directory and root for a model to be evaluated is defined and a related set of data is needed. For each model, the related data can be defined by accepting the defaults, in which case nothing else needs to be included in the file. Alternatively, the data can be listed in the Model_Paths input block or the Model_Groups input block. These options are described below in the sections on each of these input blocks.

Blockformat KEYWORDS

If blockformat is specified as KEYWORDS, blockbody is expected to be a series of phrases of the form keyword=value. For example, the Output_Control input block described later in this chapter has a keyword named WritePreds. If blockformat KEYWORDS is specified, the WritePreds keyword would be defined using a phrase such as WritePreds=yes. There can be spaces on each side of the equal sign. Phrases can occur on separate lines or can occur on the same line if they are separated by spaces. No other type of delimiter is recognized.

Here is a simple example input block using blockformat keywords. The keywords are defined later in this chapter in the section on the Options input block.

```
BEGIN Output Control Keywords
#With blockformat keywords, comments can be placed to the right of the
#data
WritePreds=yes        ##write files of predictions and variances
WriteParamNative=yes  ##write files of native parameter values&variances
WriteParamRegress=yes ##write files of regression parameters&variances
END Output Control
```

Blockformat TABLE

If blockformat is specified as TABLE, the first non-comment line of blockbody is in the format:

NROW=nr NCOL=nc [COLUMNLABELS] [DATAFILES=$nfiles$] [GROUPNAME=$gpname$]

The format of the rest of the blockbody depends on whether DATAFILES is listed, as shown in Table 6.

Table 6. For blockformat TABLE, the format of blockbody after the first line with the optional keyword COLUMNLABELS both without and with the optional keyword DATAFILES.

[Each line can contain up to 2,000 characters]

Without DATAFILES keyword[1]				With DATAFILES keyword[2]	
[column-name column-name...]				[column-name column-name...]	
val	val	...	val	pathname	[SKIP=nskip]
val	val	...	val	pathname	[SKIP=nskip]
...				...	

[1] The number of lines needed equals the value defined by NROW=nr, excluding the optional line containing column-names.

[2] The number of lines needed equals the value defined by DATAFILES=$nfiles$, excluding the optional line containing column-names. The value defined for NROW is not used.

Definition of keywords and variables:

NROW and **NCOL** are required keywords.

nr is the number of rows in the table of values to be entered.

nc is the number of columns in the table of values to be entered.

COLUMNLABELS is an optional keyword.

> **COLUMNLABELS** omitted: A default column order is used to identify the data in the columns of the table. Default column orders are only available for *blocklabel*s as identified in Table 4. If a default column order is not available COLUMNLABELS is required.

> **COLUMNLABELS** listed: Column names are used to identify the data in the columns of the table. Data are read for columns with column names that are equivalent to keywords defined for this *blocklabel*. The keywords for each MMA input block are defined later in this chapter. Data in columns with other labels are ignored. This allows data sets to contain columns that are not used by MMA. However, it also means that misspelled keywords are not flagged as errors. Default values are used if keywords are misspelled.

DATAFILES is an optional keyword.

> **DATAFILES** omitted: *nr* rows of data are read. The data type expected for **val** depends on the *blocklabel* and on *column-name* if the optional keyword **COLUMNLABELS** is listed. All data values for a row need to be on one line of the file.

> **DATAFILES** listed: A list of file pathnames is read, as shown in the second column of Table 6. The number of pathnames read equals **nfiles**, for example, if **DATAFILES**=2, two pathnames are read. Each **pathname** is the path to a file from which rows of data are read. Paths with spaces need to be enclosed in double quotes. Each file needs to contain rows of data in columns in either the default column order or the order defined by the **column-name**

entries, if specified. Data read from all files are combined as if read from one file. Each file is read in order until *nr* rows of data have been read. If **SKIP=***nskip* is specified, *nskip* lines at the beginning of the file are ignored, and reading of data starts on the following line.

Each line can contain up to 2,000 characters.

GROUPNAME is an optional keyword.

For blocks that use groups, **GROUPNAME=***gpname* can be used to assign a group name to all rows in the table. *gpname* is the name of the group. If **GROUPNAME=***gpname* is present, **GROUPNAME** is omitted from the default list of columns and can not be included with the COLUMNLABELS option.

Here is a simple example using *blockformat* TABLE and two keywords of the Model_Paths input block, `PathAndRoot` and `GroupName` (MMA input blocks are listed in Table 4 and described later in this chapter).

```
BEGIN MODEL PATHS TABLE
nrow=7  ncol=2 columnlabels
PathAndRoot       GroupName
..\DATA\Z1\1\Z     Z1
..\DATA\Z2\1\Z     Z2
..\DATA\Z2\2\Z     Z2
..\DATA\Z2\3\Z     Z2
..\DATA\Z3\1\Z     Z3
..\DATA\Z3\2\Z     Z3
..\DATA\Z3\3\Z     Z3
END MODEL PATHS
```

The example above might result, for instance, if there are three groups of models that are to be considered separately because they estimate different numbers of parameters (for example, perhaps 1, 2, and 3 parameters for groups Z1, Z2, and Z3, respectively). For Z1, there is only one model because the property of interest is homogeneous throughout the model domain. For the other two groups, there may be a number of different models that use different procedures for distributing the estimated parameters in space. In this example we indicate three models in each of groups Z2 and Z3.

If all of the models were in the same group, the input block might appear as follows:

```
BEGIN MODEL PATHS TABLE
nrow=7  ncol=1 columnlabels GroupName=ZSameGroup
PathAndRoot
..\DATA\Z1\1\Z
..\DATA\Z2\1\Z
..\DATA\Z2\2\Z
..\DATA\Z2\3\Z
..\DATA\Z3\1\Z
..\DATA\Z3\2\Z
..\DATA\Z3\3\Z
END MODEL PATHS
```

Blockformat FILES

If **blockformat** is specified as **FILES**, the input block can contain one or more lines, each containing a pathname to a file. Lines with # as the first character are interpreted as comments and are ignored. The data in each file need to be in an input block with a **BEGIN** statement and an **END** statement.

Here is a simple example input block that would appear in the main MMA input file using the **blockformat** FILES option. This input block indicates that files named MODEL-LIST-1 and MODEL-LIST-2 are to be read.

```
BEGIN MODEL PATHS FILES
MODEL-LIST-1
MODEL-LIST-2
END MODEL_PATHS
```

File MODEL-LIST-1 might contain the following information (the keywords PathAndRoot and GroupName are defined later in the Model_Paths input block instructions).

```
BEGIN MODEL PATHS TABLE
nrow=4   ncol=2 columnlabels
PathAndRoot         GroupName
..\DATA\Z2\1\Z      Z1
..\DATA\Z2\1\Z      Z2
..\DATA\Z2\2\Z      Z2
..\DATA\Z2\3\Z      Z2
END MODEL PATHS
```

and File MODEL-LIST-2 might contain the following information.

```
BEGIN MODEL PATHS TABLE
nrow=3   ncol=1 columnlabels GroupName=z3
PathAndRoot
..\DATA\Z3\1\Z
..\DATA\Z3\2\Z
..\DATA\Z3\3\Z
END MODEL PATHS
```

The Two Alternatives for Reading Data from Other Files

The discussion above shows that data can be read from other files using either of the *blockformat* alternatives TABLE or FILES. The mechanisms and their characteristics are summarized in Table 7.

Table 7. Alternatives for reading data from files.

Blockformat TABLE with DATAFILES	Blockformat FILES
There is only one Begin blockformat and End blockformat block.	There are Begin blockformat and End blockformat statements in the main input files and in each of the listed files.
All data are read as a table.	Blockformat can change based on the designations in the Begin statements.

Using *blockformat* TABLE in the last example would result in the following input block in the main MMA input file.

```
BEGIN MODEL PATHS TABLE
nrow=7  ncol=2 DataFiles=2 ColumnLabels
PathAndRoot   Groupname
MODEL-LIST-1
MODEL-LIST-2
END MODEL PATHS
```

The files MODEL_LIST_1 and MODEL_LIST_2 would only include rows and columns of data and would both need to include the Groupname as follows:

File MODEL-LIST-1 might contain the following information,

```
..\DATA\Z2\1\Z     Z1
..\DATA\Z2\1\Z     Z2
..\DATA\Z2\2\Z     Z2
..\DATA\Z2\3\Z     Z2
```

and File MODEL-LIST-2 might contain the following information.

```
..\DATA\Z3\1\Z     Z3
..\DATA\Z3\2\Z     Z3
..\DATA\Z3\3\Z     Z3
```

Options Input Block (optional)

The Options input block is the first MMA input block listed in Table 4.

The Options input block can be used to control the information written to the main output file. More information often is useful for finding errors in input files.

One keyword can be read from the Options input block. Usually, the KEYWORDS blockformat is used.

Verbose   - Flag that controls what is written to the MMA main output file as follows. The default is Verbose=3 to provide information for new applications and users, but Verbose=0 is suggested for most circumstances.

Verbose	Output
0	No extraneous output.
1	Warnings.
2	Warnings, notes.
3 (default)	Warnings, notes, echo selected input.
4	Warnings, notes, echo all input.
5	Warnings, notes, echo all input, plus some miscellaneous information. Includes all values read from process-model output files.

An example Options input block is as follows.

```
BEGIN OPTION KEYWORDS
VERBOSE=0
END OPTION
```

Output_Control Input Block (optional)

The Output_Control input block allows the user to control printing of six files that can be useful for debugging. Keywords are as follows:

WritePreds — yes: print two files, one containing the values of predictions to be averaged, and the other containing their variances. no: do not print these files. In chapter 6, the files are referred to as MMAroot._IndividPred and MMAroot._IndividPredVar. The default is WritePreds=no.

WriteParamNative — yes: print two files, one containing the native values of the parameters to be averaged for each model and the other containing their variances in native space. "Native" refers to the values that are not log-transformed, and is important only for log-transformed parameters. no: do not print these files. In chapter 6, the files are referred to as MMAroot._IndividParamNative and MMAroot._IndividParVarNative. The default is WriteParamNative=no.

WriteParamRegress — yes: print two files. One contains regression values of the parameters to be averaged for each model. "Regression" means that log-transformed values are used for log-transformed parameters; values for other parameters remain in native space. The other file contains parameter variances in regression space. no: do not print these files. In chapter 6, the files are referred to as MMAroot._IndividParamRegress and MMAroot._IndividParVarRegress. The default is WriteParamRegress=no.

Regression space differs from native space if any parameters are log-transformed. In native space, the untransformed values are always reported. In regression space, log10 values are reported for log-transformed parameters. The values of concern here are parameter values and variances.

The models analyzed by MMA may be a subset of the models listed in the MMA main input file. See the section in chapter 2 entitled "Omitting selected models from the analysis" for additional information.

Default columns are not defined for this block.

An example Output_Control input block is as follows.

```
BEGIN OUTPUT CONTROL KEYWORDS
WRITEPREDS=YES
END OUTPUT CONTROL
```

Model_Groups Input Block (optional)

The Model_Groups input block lists names of model groups and whether to compute model-averaged parameters.

Keywords are as follows:

GroupName   - Name of a group of models that will be treated similarly with respect to reasonable parameter values and model averaging. The default is GroupName=Default.

Avg   - yes: compute model-averaged parameters for this group. no: do not computer model-averaged parameters for this group. The default is Avg=no.

If blockformat KEYWORDS is selected by designation or default, keywords associated with a model group in the Model_Groups input block need to be grouped together and follow the GroupName keyword. The GroupName keyword needs to be the first keyword on a new line. GroupName and associated keywords are repeated to define multiple groups.

Default columns are not defined for this block.

An example Model_Groups input block is as follows, where Z1, Z2, and Z3 are example groupnames.

```
BEGIN MODEL_GROUPS TABLE
NROW=3 NCOL=2 COLUMNLABELS
GROUPNAME     AVG
Z1              NO
Z2             YES
Z3             YES
END MODEL GROUPS
```

Groupnames can be used in three input blocks: Param_Eqns, Param_Avgs, and Model_Paths. The models included in each group are defined in the Model_Paths input block.

Param_Eqns Input Block (optional)

The Param_Eqns input block associates equations describing acceptable parameter-value relations with model groups.

The acceptable parameter-value relations may define relative relations of parameters such as: the value of param1 is expected to be less than the value of parameter param2 ("param1 .lt. param2"); or individual conditions, for example: param1 is expected to be less than a certain value (for example, "param1 .lt. 1.0"). Operators, functions, and logical operators that can be used in the equation are listed in the section "Equation Protocols" later in this chapter.

If no parameter equations are defined, all parameter values are classified as being acceptable.

Keywords are as follows:

ParEqnName   - Name of a parameter-relation equation. The name must start with a letter and can include up to 20 alphanumeric characters (A–Z, a–z, 0–9) and underscore (_). The name is not case sensitive. Each name needs to be unique.

ParEqn   - Equation defining acceptable parameter values. Each equation may include the names of parameters estimated by every model in the group of models considered, any number of arithmetic operators and functions, and one logical operator. The possible arithmetic operators and functions and logical operators are listed in the section "Equation Protocols." The default is no equation.

GroupName   - Name of a group of models that will be treated similarly with respect to reasonable parameter values. The default is GroupName=Default.

The default column order for this block is the order in which the keywords are defined above.

An example Param_Eqns input block follows, using K1, K2, and K3 as example parameter names. The syntax used for the ParEqn expressions is described in the "Equation Protocols" section of this chapter.

```
BEGIN PARAM EQNS TABLE
NROW=5 NCOL=3 COLUMNLABELS
ParEqnName     GroupName      ParEqn
KorderZ2 1 2    Z2           K1.lt.K2
KvalueZ2low     Z2           "K1 .gt. 0.005"
KvalueZ2high    Z2           K2.lt.0.02
KorderZ3 1 2    Z3           K1.lt.K2
KorderZ3 2 3    Z3           K2.lt.K3
END PARAM EQNS
```

This example indicates that the following conditions need to be satisfied for a model to be included in the analysis conducted by MMA:

For models in group Z2, the estimated value of parameter K1 needs to be less than that of K2, the value of K1 needs to be greater than 0.005, and K2 needs to be less than 0.02.

For models in group Z3, the estimated value of parameter K1 needs to be less than that of K2, and the value of K2 needs to be less than that of K3.

If expressions include blank space they need to be in double quotes. This and other details are explained in the section "Equation Protocols" later in this chapter.

Param_Avgs Input Block (optional)

The Param_Avgs input block associates parameters to be averaged with group names.

It is important to recognize that obtaining meaningful average parameters requires that the parameter name represents the same entity in each model. This is discussed further in chapter 2 in the section "Including prior information on parameters."

Keywords are as follows:

ParAvgName   - Name of parameter to be averaged. There is no default.

GroupName   - Name of a group of models to involve in the averaging. The parameter needs to have the same name and be estimated for every model in the group, or MMA will stop with an error message. Listed models are not necessarily included in the analysis; see the section in chapter 2 entitled "Omitting selected models from the analysis." For the model-averaged value to have meaning, the parameter needs to represent the same entity in every model in the group, as described in the section "Including prior information on parameters." The default is GroupName=Default.

Avg   - yes: calculate model-averaged values for this parameter. no: do not calculate model-averaged values. The default is Avg=no.

If blockformat KEYWORDS is selected by designation or default, keywords associated with a parameter in the Param_Avgs input block need to be grouped together and follow the ParAvgName keyword. The ParAvgName keyword needs to be the first keyword on a new line. To list multiple parameters, repeat ParAvgName and associated keywords.

The default column order for this block is the order in which the keywords are defined above.

An example PARAM_AVGS input block is as follows.

```
BEGIN PARAM AVGS TABLE
NROW=5 NCOL=3 COLUMNLABELS
ParAvgName      GROUPNAME      AVG
K1              Z2             YES
K2              Z2             YES
K1              Z3             YES
K2              Z3             YES
K3              Z3             YES
END PARAM AVGS
```

Model_Paths Input Block (required)

The Model_Paths input block lists the paths (including the root file name) to models that will be evaluated.

Keywords are as follows:

PathAndRoot — Computer directory path and root of the output files of the models being evaluated using MMA. The path can be relative or absolute. The root needs to match the filename prefix of the data-exchange files to be read by MMA. When using UCODE_2005 to produce these files, this is the filename prefix defined on the UCODE_2005 command line. There is no default.

PriorModProb — Prior probability for this model. The default is PriorModProb=1/(number of models). Prior model probabilities are proportionally adjusted when models are omitted (see the section in chapter 2 entitled "Omitting selected models from the analysis").

GroupName — Group name associated with this model. Groupnames are defined in the Model_Groups input block. The default is Groupname=Default.

If blockformat KEYWORDS is selected by designation or default, keywords associated with a path in the Model_Paths input block need to be grouped together and follow the PathAndRoot keyword. The PathAndRoot keyword needs to be the first keyword on a new line. PathAndRoot and associated keywords are repeated to list multiple paths.

The default column order for this block is the order in which the keywords are defined above.

An example Model_Paths input block is as follows.

```
BEGIN MODEL PATHS TABLE
# USE DEFAULT FOR PriorModProb
nrow=7  ncol=2 columnlabels
PathAndRoot      GroupName
..\DATA\Z1\1\Z     Z1
..\DATA\Z2\1\Z     Z2
..\DATA\Z2\2\Z     Z2
..\DATA\Z2\3\Z     Z2
..\DATA\Z3\1\Z     Z3
..\DATA\Z3\2\Z     Z3
..\DATA\Z3\3\Z     Z3
END MODEL PATHS
```

Preds Input Block (optional)

The Preds input block lists the name of predictions to be model averaged. There is only one keyword for this input block. The prediction names listed need to match the names of predictions produced in the model data-exchange files with filename extension _linp. Each prediction to be averaged needs to exist for every model in the evaluation. Some models may be omitted from the analysis, as described in the chapter 2 section entitled "Omitting selected models from the analysis."

Keywords are as follows:

Prediction   - Name of a prediction to be considered by MMA. There is no default.

Default column order for this block is: Prediction

An example Preds input block is as follows.

```
BEGIN PREDS TABLE
NROW=5 NCOL=1 COLUMNLABELS
PREDICTION
PRED1
PRED2
PRED3
PRED4
PRED5
END PREDS
```

Analyses Input Block (optional)

The Analyses input block consists of one or more analysis names, each associated with an equation defining the criterion by which the models are ranked and an equation used to calculate the posterior model probabilities.

If this input block is not included in the MMA main input file, the default is to produce analyses for four model measures. The four measures depend on whether or not prior information equations are to be included. If prior information equations are included, the four default model measures are AICwPri, AICcwPri, BICwPri, and KICwPri. If the models do not include prior information equations on parameters or the names of the prior information equations differ between models, then the four default model measures are AICObs, AICcObs, BICObs, and KICObs. The names of the prior information equations for parameters are read for each model from the data-exchange file with filename extension _pr.

If analyses are specified in the Analyses input block, they are produced instead of the default analyses. To obtain the default measures as well, they would need to be listed explicitly in the Analyses input block.

Keywords are as follows:

AnalysisLabel   - Label used in the name of the files containing results from the MMA run. The default is to conduct four analyses with
AnalysisLabel=AICwPri, AICcwPri, BICwPri, and KICwPri
or, if parameter prior information equations can not be used, as discussed in the section "Prior information on parameters" in chapter 2,
AnalysisLabel =AICObs, AICcObs, BICObs, and KICObs.

     The analysis label used does not affect the analysis conducted. It is used to name an MMA output file and to label values written in MMA output files, as discussed in chapter 6.

CritEqn   - An equation defining the criterion for the associated analysis label and used to calculate posterior model probabilities and associated statistics, by which the models are ranked. The default is to conduct four analyses with
CritEqn=AICwPri, AICcwPri, BICwPri, and KICwPri
or, if prior information equations can not be used as discussed in the section "Prior information on parameters" in chapter 2,
CritEqn=AICObs, AICcObs, BICObs, and KICObs.

     CritEqn can include any label from the left-most column of Table 8, any of the three bold expressions from the leftmost column of Table 9 (DISTCNTW, DISTCNT_SGNW, and DIST_CNTMGW), and the operators and functions listed in Table 11. Commonly used items from Table 8 include: CEV*, MLOF*, AIC*, AICc*, BIC*, and KIC*, where here the "*" is replaced by "obs" or "wPri" as discussed in chapter 3. To construct more complicated equations, see the section "Equation Protocols."

PrEqn
    - Numerator of the posterior model probability equation (eq. 2.12a or b) to be used for this analysis. The normalization accomplished by the denominator of equations 2.12a or b is done internally by MMA. PrEqn can include any of the expressions listed in Table 10 and the operators and functions from Table 11. To construct more complicated equations, see the section "Equation Protocols." The default is

    PrEqn="exp(-0.5*(valcrit-mincrit))*PriorModProb".

PriorModProb is defined in the Model_Paths input block. As discussed there, if values are not assigned, PriorModProb is set to the same value for all models. This makes it cancel out of the calculation of posterior model probabilities, resulting in the default being equivalent to equation 2.12a.

If blockformat KEYWORDS is selected by designation or default, the AnalysisLabel keyword needs to be the first keyword on a new line and keywords associated with each analysis need to be grouped together and follow the AnalysisLabel keyword. AnalysisLabel and associated keywords can be repeated to conduct multiple analyses.

The default column order for this block is the order in which the keywords are defined above.

The example Analyses input block that follows would result in models being ranked and weighted using (1) the K-L information method described in chapter 2, (2) a method based on the sum-of-weighted squared residuals (often included to display goodness of fit but not usually suitable for model selection because it lacks a penalty for added parameters), and (3) maximum likelihood Bayesian model averaging. All include prior model probabilities. Analyses(1) o (3) use criteria that include prior information on the parameters; analyses (4) to (6) include only observations. Analysis (7) uses the SWSR to assign posterior model probabilities in a linear manner with respect to SWSR and ignores the prior model probability.

```
BEGIN ANALYSES TABLE
#Quotes are needed if there are any spaces in the equation
NROW=7 NCOL=3 COLUMNLABELS
AnalysisLabel CritEqn     PrEqn
K-L           AICcwPri "exp(-0.5*(valcrit-mincrit)) * PriorModProb"
SWSR          SWSRwPri "exp(-0.5*(valcrit-mincrit)) * PriorModProb"
MLBMA         KICwPri  "exp(-0.5*(valcrit-mincrit)) * PriorModProb"
K-L-OBS       AICcObs   exp(-0.5*(valcrit-mincrit))*PriorModProb
SWSR-OBS      SWSRObs   exp(-0.5*(valcrit-mincrit))*PriorModProb
MLBMA-OBS     KICObs    exp(-0.5*(valcrit-mincrit))*PriorModProb
SWSR          SWSRwPri 1.+((mincrit-valcrit)/(maxcrit-mincrit))
END ANALYSES
```

Analysis-specific output files with filenames formed in part by the statistic labels are produced as described in chapter 6.

Table 8. Variables and measures of model fit to observations that can be used in the model criterion defined with the Analyses input block. For additional information, see chapters 2 and 3. [R^2, the correlation between the two data sets listed. The statistics are reported in filenames constructed as MMAroot followed by a period and the characters in parentheses. MMAroot is defined on the MMA command line, see chapter 4]

Variable[1]	Description; also see chapter 3	
NPE	Number of estimated parameters	
NOBS	Number of observations	
NPR	Number of prior estimates	

Measure[1,2]	Description; also see chapter 3 and listed equations	Preferred value
Overall measures of model fit (_mma)		
SWSR*	Sum of weighted squared residuals; eq. 2.1b	small
CEV*	Calculated Error Variance	1.0
MLOF*	Maximum Likelihood Objective Function	small
AIC*	Akaike's Information Criterion; eq. 2.1	small
AICc*	Akaike's Modified Information Criterion; eq. 2.2a	small
BIC*	Bayesian Information Criterion; eq. 2.7	small
KIC*	Kashyap's Criterion; eq. 2.8	small
XTwX*	Natural log of determinant of the matrix $\underline{X}^T \underline{\omega} \underline{X}$	--
Statistics from graphs of observed values and simulated equivalents (_mma_gstats)		
R2_os*	R^2 for the line that provides the best fit	1.0
Int_os*	Intercept of line that provides the best fit	0.0
Slp_os*	Slope of line that provides the best fit	1.0
Statistics from graphs of weighted residuals and simulated equivalents (_mma_gstats)		
R2_ws*	R^2 for the line that provides the best fit	0.0
Int_ws*	Intercept of line that provides the best fit	0.0
Slp_ws*	Slope of line that provides the best fit	0.0
Statistics from graphs of weighted observed values and weighted simulated equivalents (_mma_gstats)		
R2_ww*	R^2 for the line that provides the best fit	1.0
Int_ww*	Intercept of line that provides the best fit	0.0
Slp_ww*	Slope of line that provides the best fit	1.0
Statistics that test if weighted residuals are distributed normally and independently (_mma_gstats)		
R2_nm*	R^2 between weighted residuals and standard normal deviates	1.0
Statistics from graphs of residuals and time (_mma_xyzt)		
R2_rt	R^2 for the line that provides the best fit	0.0
Int_rt	Intercept of line that provides the best fit	0.0
Slp_rt	Slope of line that provides the best fit	0.0
Statistics from graphs of weighted residuals and time (_mma_xyzt)		
R2_wrt	R^2 for the line that provides the best fit	0.0
Int_wrt	Intercept of line that provides the best fit	0.0
Slp_wrt	Slope of line that provides the best fit	0.0

[1] Labels used in MMA output files and in equations defined in the Analyses input block.
[2] In the equations of the Analyses input block, the "*" can be replaced by "Obs" or "wPri". In the MMA output files, "*" can be replaced by "Obs", "wPri", "PR-O", "_%", and "_Chg". See chapter 3 for additional information and chapter 5 for file headers.

Table 9. Statistics calculated for each model that measure how randomly the weighted residuals are distributed in space.
[i, coordinate direction that can be set to X, Y, or Z, where the axes are consistent with the axes used to determine the values in the MMAroot.xyzt input file described in chapter 5. The statistics in bold are expressions that can be used in equations defined in the Analyses input block. The values of these statistics for each model are listed in output file MMAroot._mma_xyzt]

Column label	Description	Preferred values
CNT_LOCi	Centroid of the i coordinate of the locations of observations as listed in the MMAroot.xyzt file. Calculated by summing all coordinate values for the i direction and dividing by the number of observations.	
DIFCNT_Wi	Measures whether the weighted residuals are distributed randomly in each coordinate direction. Calculated as the difference between the centroid of the weighted residuals calculated for the i direction and CNT_LOCi. The centroid of the weighted residuals in the i direction is calculated by summing the products of the weighted residuals times the i coordinate and dividing by the number of observations.	
DISTCNTW	Measures whether the weighted residuals are distributed randomly in space. Calculated as the square root of the sum over i of the squared DIFCNT_Wi.	0.0
DIFCNT_SGNWi **DISTCNT_SGNW**	Measures whether positives and negatives are randomly distributed in space regardless of magnitude. Calculated similarly to DIFCNT_Wi and DISTCNTW except negative weighted residuals are replaced by -1, positive weighted residuals are replaced by +1.	0.0
DIFCNT_MGWi **DIST_CNTMGW**	Measures whether magnitudes of the weighted residuals are randomly distributed in space regardless of sign. Calculated similarly to DIFCNT_Wi and DISTCNTW except the absolute values of the weighted residuals are used.	0.0

Table 10. Additional expressions that can be used in PrEqn.
[Criterion values are defined by the CritEqn keyword in the Analyses input block;
models are included if (a) parameter estimation converged and (b) parameter values meet
restrictions defined in the Param_Eqns input block]

Expression	Description
Expressions that produce one value for all included models	
MinCrit	minimum criterion value
MaxCrit	maximum criterion value
SumCrit	sum of criterion values
AvgCrit	average of criterion values
Expressions that produce one value for each model	
ValCrit	criterion value for each model
PriorModProb	Defined as described for the Model Paths input block.

Equation Protocols

In MMA, equations can be defined in two input blocks: Param_Eqns, and Analyses. Equations are given a name and a mathematical expression. The mathematical expressions available for equations defined in the Analyses input block differ from those available for equations defined in the Param_Eqns input block.

In the Analyses block for keywords CritEqn and PrEqn, the mathematical expression represents what would normally be placed on the right side of an "=" sign; the equal sign is not included in the expression. For CritEqn, the expression consists of arithmetic operators and functions of Table 11, model-measure labels from Table 8 and those listed in bold in Table 9, and constants. Acceptable arithmetic operators and functions are listed in Table 11. For PrEqn, the expression consists of arithmetic operators and functions of Table 11, functions of the criterion values from Table 10, and constants.

In the Param_Eqns block for keyword ParEqn, the logical operators listed in Table 12 also can be used in conjunction with arithmetic operators and functions of Table 11, parameter names (read from data-exchange file _pc in the directories containing the results of the model regressions), and constants. In this situation, the equation is a logical statement such as "K1 < 0.5*K2".

The order in which mathematical operations are carried out in evaluating a mathematical expression is the same as that used in normal mathematical operations, that is: raising to a power, followed by multiplication and division, followed by unary addition and subtraction, followed by binary addition and subtraction. Parentheses can be used to override or clarify this order.

Table 11. Arithmetic operators and functions available for equations.

Arithmetic operator	Operation
** or ^	Power. $a**b$ or a^b is interpreted as "a raised to the power b."
/	Division. a/b is interpreted as "a divided by b."
*	Multiplication. $a*b$ is interpreted as "a multiplied by b."
−	Subtraction. This can be a unary or binary operator. $a–b$ is interpreted as "a minus b"; $–a$ is interpreted as "negative a."
+	Addition. This can be a unary or binary operator. $a+b$ is interpreted as "a plus b"; $+a$ is interpreted as "positive a."
( )	Parentheses. Terms within parentheses are evaluated first. For example: $5 + 4 * 3$ is evaluated as 17. However $(5 + 4) * 3$ is evaluated as 27.

Function	Definition
abs( )	Absolute value. Argument can be any floating-point number.
cos( )	Cosine. Argument can be any floating-point number supplied in radians.
acos( )	Inverse cosine. Absolute value of argument must be between -1 and 1. Value is returned in radians.
sin( )	Sine. Argument can be any floating-point number supplied in radians.
asin( )	Inverse sine. Absolute value of argument must be between -1 and 1. Value is returned in radians.
tan( )	Tan. Argument can be any floating-point number supplied in radians.
atan( )	Inverse tan. Argument can be any floating-point number. Value is returned in radians.
cosh( )	Hyperbolic cosine. Argument can be any floating-point number.
sinh( )	Hyperbolic sine. Argument can be any floating-point number.
tanh( )	Hyperbolic tan. Argument can be any floating-point number.
exp( )	Exponential. Argument can be any floating-point number.
log( )	Log to base e. Argument must be a positive floating-point number.
log10( )	Log to base 10. Argument must be a positive floating-point number.
sqrt( )	Square root. Argument must be non-negative.
min(, ,)	Minimum of a series of numbers. Arguments can be any set of floating-point numbers.
max(, ,)	Maximum of a series of numbers. Arguments can be any set of floating-point numbers.
mod(,)	Remainder. $mod(a,b)$ is the remainder after a is divided by b.

Table 12. Logical operators available for equations.

Logical operator	Operation
.lt.	<u>Less than</u>. *a*.lt.*b* is *true* if *a* is less than *b*.
.le.	<u>Less than or equal to</u>. *a*.le.*b* is *true* if *a* is less than or equal to *b*.
.eq.	<u>Equal to</u>. *a*.eq.*b* is *true* if *a* equals *b*.
.gt.	<u>Greater than</u>. *a*.gt.*b* is *true* if *a* is greater than *b*.
.ge.	<u>Greater than or equal to</u>. *a*.ge.*b* is *true* if *a* is greater than or equal to *b*.
.ne.	<u>Not equal to</u>. *a*.ne.*b* is *true* if *a* does not equal *b*.
.and.	<u>And</u>. *a*.and.*b* is *true* if both *a* and *b* are true; for example ((1.lt.10).and.(6.lt.7)) is *true*.
.or.	<u>Or</u>. *a*.or.*b* is *true* if *a* is *true* or *b* is *true* or both are *true*; for example ((1.lt.10).or.(1.lt.0)) is *true*.

The following are some examples of acceptable equations. It is assumed that values are available for variables stat1 and stat2.

```
"stat1 + sqrt(stat2*stat1)"

sqrt(CEV)

'exp(3.0 * sqrt(stat1/stat2))'

stat1

1.0

K1.lt.K2

K3.lt.10*K2
```

These examples demonstrate the following:

1. Spaces can be left between operators, variable names, brackets, and so on if the equation is enclosed in double or single quotes. However, a variable name can not include a space.

2. An equation entity that is not an operator or a function is first treated as a number. If it can not be read as a number, it is assumed to be a variable. To avoid confusion, variable names can not begin with a number.

3. If an illegal argument is supplied to any function (for example if a negative number is provided as the argument to a log or sqrt function), an error condition arises, the error is reported, and MMA execution stops.

The MMAroot.xyzt Input File Used to Define Observation Locations and Times

If an MMAroot.xyzt file exists in the folder where MMA is executed, MMA uses that file to provide the location and time of each observation to conduct the analyses described in the last two sections of chapter 3. The file needs to be named using the filename prefix defined on the MMA command line, a period, and the filename extension "xyzt." The filename prefix is referred to as MMAroot in this document, which is why the file is referred to as MMAroot.xyzt.

The criteria that are related to space and time are printed in the MMAroot._mma_xyzt file and some of the criteria may be selected by the user as part of the CritEqn. (For more information see Table 9.)

If the MMAroot.xyzt file is not found, then spatial and temporal criteria that require the missing information are assigned a value of $1.0 \times 10^{+30}$. The large value is used when the correct information is not available so that the results are obviously meaningless.

The MMAroot.xyzt input file is constructed like the fn.xyzt input file described for UCODE_2005 (Poeter and others, 2005, p. 146), and fn.xyzt files constructed for UCODE_2005 can be used for MMA. The input instructions are as follows.

The MMAroot.xyzt file has the following characteristics. The first line is ignored and can be used to list column headings, projection information, or other information. The rest of the file needs to be composed of lines containing five columns of data:

OBSERVATION NAME   X   Y   Z   TIME

Items may be separated by spaces, tabs, or commas. As long as at least one space, comma, or tab follows the time, then additional data or comments can be included to the right on the same line and are ignored by MMA.

All observations used in the alternative models need to appear in the MMAroot.xyzt input files used for MMA. As mentioned in chapter 1 and elsewhere in this report, all alternative models need to use the same set of observations. If some observations are not listed in the MMAroot.xyzt file, then MMA will terminate with an error message. Observations are ignored if they are listed in the MMAroot.xyzt file and are not used in the regressions being evaluated by MMA.

Chapter 6: MMA Output

Files output by MMA provide the information needed for model analysis. The user controls which files are produced using the Output_Control input block, or by declining to request an option such as parameter or prediction averaging. All output files begin with the root name described on the program command line for MMA (represented as "MMAroot" here) and end with an MMA-defined extension, as shown below.

Except for the MMAroot.#mout file, all of the files are constructed using the conventions of JUPITER API data-exchange files, and therefore are suitable for importing to a spreadsheet to create graphs for model analysis and for reports, or to be read easily by other computer programs.

Files that are Always Produced

MMA always produces one main output file, seven data-exchange files that always have the same filename extension, and one data-exchange file for each analysis conducted. These files are produced even when the only input block defined is the Model_Paths input block.

Main Output File MMAroot.#mout

The main input file is designed to be read by MMA users.

MMAroot.#mout   - Summarizes tasks undertaken by MMA and contains warning and error messages.

If VERBOSE of the Options input block is greater than zero by designation or default, the additional output is written to this file.

Seven Data-Exchange Files that Always Have the Same Filename Extension

The first six data-exchange files listed below contain a column for each measure listed in Table 8 and bold items of table 9 followed by a column that identifies the associated model. The model-measure labels listed in Table 8 and 9 are included as column headers enclosed in double quotes. All models are listed whether or not they meet the criteria for inclusion in the analysis.

In the three output files listed below, a value of $1.0 \times 10^{+30}$ is assigned for models that did not converge or for which there are unreasonable parameter values, as defined using the Param_Eqns input block.

MMAroot._mma   - Lists the value of each overall measure of model fit described in chapters 2 and 3 and listed in Table 8.

This file includes initial columns with the model number, model name, NPE, NOBS, and NPR, and a final column with the path to the model. See chapter 3 or Table 8 for definitions of NPE, NOBS, and NPR.

MMAroot._mma_gstats   -Lists the value of each model measure derived from graphs that compare two quantities, as described in chapter 3 and listed in Table 8.

This file includes initial columns with the model number and model name, and a final column with the path to the model.

MMAroot._mma_xyzt   - Lists the value of each model measure of bias in time and space, as described in chapter 3 and listed in Table 8 and Table 9.

This file includes initial columns with the model number and model name, and a final column with the path to the model.

If an MMAroot.xyzt file described in chapter 5 is not present in the folder where MMA is executed, this file contains the message: "XYZT data are not available, XYZT measures are not written."

The following comments apply to the next three output files, which provide ranks for the models.

1. Some model measures are not appropriate for ranking. In the associated columns, all models are assigned a value of zero.

2. For each column with model ranks, the best model based on that measure is ranked 1, where best is determined using the preferred values listed in Table 8 and Table 9. If the values are equal for two or more models, the rank is repeated and an appropriate number of subsequent ranks are not assigned.

3. All models that did not converge or do not meet the reasonable parameter specifications defined in the Param_Eqns input block are assigned the lowest rank The lowest rant equals one plus the total number of models that converged and met the reasonable parameter specifications defined in the Param_Eqns input block. That is, if there are 10 models and 5 converged and met the reasonable parameter specifications, the other models would be assigned a rank of 6.

MMAroot._rank   - Lists the rank of each model based on the overall measures of model fit described in chapters 2 and 3 and listed in Table 8.

This file includes initial columns with the model number, model name, NPR, NOBS, and NPR, and a final column with the path to the model. See chapter 3 or Table 8 for definitions of NPE, NOBS, and NPR.

MMAroot._rank_gstats   - Lists the rank of each model based on measures derived from graphs that compare two quantities, as described in chapter 3 and listed in Table 8.

This file includes initial columns with the model number and model name, and a final column with the path to the model.

MMAroot._rank_xyzt   - Lists the rank of each model for measures of bias in time and space, as described in chapter 3 and listed in Table 8 and Table 9.

This file includes initial columns with the model number and name, and a final column with the path to the model.

If an MMAroot.xyzt file described in chapter 5 is not present in the folder where MMA is executed, this file contains the message: "XYZT data are not available, XYZT measures are not written."

The seventh file simply provides a list of the names and the paths and roots of the analyzed models.

MMAroot._ModelNamesPaths - Lists the names and the paths and roots of the models used in the analysis (those that converged and met the reasonable parameter specifications defined in the Param_Eqns input block).

Data-Exchange Files with Filename Extensions that Include the Analysis Label

The following filename includes "MMAroot" and "AnalysisName", neither of which is used literally. "MMAroot" is replaced by the MMA filename root defined on the command line, as discussed in chapter 4. "AnalysisName" is replaced by default names or by names provided by the user in the Analyses input block (see chapter 5). For example, if MMAroot is defined as "z"and Anaylsis=K-L, the resulting file name is "z._anals_K-L".

One file is produced for each analysis conducted. If no Analyses input block is defined, four default analyses are conducted, as discussed in chapter 2. In this situation, "AnalysisName" is replaced by AICwPri, AICcwPri, BICwPri, and KICwPri if all models have the same prior equation names, or otherwise by AICObs, AICcObs, BICObs, and KICObs. See the section "Including priot information onparameters" in chapter 2 for additional information. The use of wPri and Obs in analysis names is discussed more in chapter 3.

If the same analysis name is used more than once, the previous file is overwritten, so define unique analysis names.

MMAroot._anals_AnalysisName   - Results of the analysis identified by "AnalysisName".

Lists the model name, prior model probability (all equal, if not specified; or renormalized if models are omitted), value of the criterion, rank, posterior model probability, delta, evidence ratio, inverse evidence ratio as a percent, and the PathandRoot from the Model_Paths input block.

Results are listed for all included models. Models can be omitted from the analysis as discussed in the section of chapter 2 entitled "Omitting selected models from the analysis." The first line of each file provides the AnalysisName and associated Criterion and Probability Equations. The second line provides column headings: Model, Prior Prob, Criterion, Rank, Probability, Delta, Evidence-Ratio, ER-Inverse as % and PathAndRoot.

Output Files Related to Predictions

Up to three output files report results related to predictions.

If model-averaged predictions are requested using the Preds input block, the following file is produced:

MMAroot._preds_AnalysisName - Model-averaged predictions, their confidence intervals, variance, name and plot symbol are provided for the group of models that converged with reasonable parameter values as defined in the Param_Eqns input block.

If requested in the Output_Control input block using keyword WritePreds (see chapter 5), the files indicated below are produced. These files make it easy to compare the predictions and predictions variances produced by the individual models, and generally are only needed to check and investigate results presented in other MMA output files.

MMAroot._IndividPred   - Lists the model names and values of each prediction (in columns) for each model (by rows) that is used in model-averaging.

MMAroot._IndividPredVar   - Lists the model names and variances of each prediction (in columns) of each model (by rows) that is used in model-averaging.

Other Output Files

There are as many as five other output files.

If parameter averaging is requested using the Param_Avgs input block, the following file is produced:

MMAroot._params_AnalysisName - Lists model-averaged parameter names, values, confidence intervals, variances and a label indicating whether the parameter was log transformed when estimated. These items are provided for each specified group of models.

Production of the following four files is controlled by keywords WriteParamNative and WriteParamRegress of the Output_Control block (see chapter 5). These files make it easy to compare the parameter values and parameter value variances produced by the individual models, and generally are only needed to check and investigate results presented in other MMA output files

The WriteParamRegress keyword controls output of the following two files.

MMAroot._IndividParamRegress - Lists the names and values of parameters for which model averages are calculated. For each group of models, the parameters are listed in columns, and each row presents results from one model.

MMAroot._IndividParVarRegress - Lists the model names and values of variance for each parameter as estimated in regression space. For each group of models, the parameters are listed in columns, and each row presents results from one model. The values listed are used in model-averaging for each group of models.

The WriteParamNative keyword controls output of the following two files.

MMAroot._IndividParamNative - Lists the model names and native estimates of each estimated parameter (in columns) for each model (in rows). These values are used in model-averaging for each group of models.

MMAroot._IndividParVarNative - Lists the model names and values of variance for each native parameter (in columns) for each model (in rows). These values are used in model-averaging for each group of models. For log-transformed parameters, variances for native parameter values are calculated as discussed by Poeter and others (2005, p. 155) and Hill and Tiedeman (2007, p. 130)

Regression space differs from native space if any parameters are log-transformed. In native space, the untransformed values are always reported. In regression space, log10 values are reported for log-transformed parameters. The values of concern here are parameter values and variances.

Entries for Omitted Models and Other Special Circumstances

As discussed in chapter 2 in the section "Overview," situations can occur in which one or more of the original set of models is omitted from analysis by MMA. In addition, it may not be possible to calculate certain statistics for all of the included models. For example, the statistics with names ending in _rt and _wrt can not be calculated if all the times have the same value. If the value of a statistic and its associated ranking can not be calculated, values are printed in the output files that clearly indicate that there is a difficulty. The values printed were mentioned as the files were described in this chapter, and also are listed in Table 13.

Table 13. Entries for statistics and rankings for which a meaningful value cannot be calculated.

File	Value	Circumstance
MMAroot._mma MMAroot._mma_gstat MMAroot._mma_xyzt	1E+30	Nonlinear regression did not converge.
		Unreasonable parameter values as determined using conditions defined in the Param_Eqns input block.
		The denominator of equation 3.1b equals zero. This equation calculates the percent change between values of measures calculated only with observations and with observations and prior information.
MMAroot._rank MMAroot._rank_gstat MMAroot._rank_xyzt	0	There is no preferred number of observations, parameters, and prior, nor a preferred value for the centroids of observation locations in the x, y, and z direction.
	Lowest rank	Nonlinear regression did not converge.
		Unreasonable parameter values as determined using conditions defined in the Param_Eqns input block.
		The denominator of equation 3.1 equals zero. This equation compares the values of statistics calculated only with observations and with observations and prior information.

Chapter 7: References

Akaike, H., 1973, Information theory as an extension of the maximum likelihood principle, *in* Petrov, B.N., ed., Second International Symposium on Information Theory: Akademiai Kiado, Budapest, p. 267–281.

Akaike, H., 1974, A new look at the statistical model identification: IEEE Transactions on Automatic Control AC, v. 19, p. 716–723.

Anderson, D.R., 2003, Multi-model inference based on Kullback-Leibler information, *in* Poeter, E., Zheng, C., Hill, M., and Doherty, J., eds., Proceedings of the conference MODFLOW and More 2003, Understanding through modeling: International Ground Water Modeling Center, Colorado School of Mines, Golden, Colorado, Sept. 16–19, 2003, p. 366–370.

Banta, E.R., Poeter, E.P., Doherty, J.E., and Hill, M.C., 2006, JUPITER: Joint Universal Parameter IdenTification and Estimation of Reliability – An application programming interface (API) for model analysis: U.S. Geological Survey Techniques and Methods, book 6, section E, chap. 1, 268 p.

Box, G.E.P., and Jenkins, G.M., 1970, Time series analysis, forecasting and control: Holden-Day, London, 553 p.

Box, G.E.P., Jenkins, G.M., and Reinsel, G.C., 1994, Time series analysis, forecasting and control, 3d edition: Prentice-Hall, Englewood Cliffs, N.J., 598 p.

Bozdogan, H., 1987, Model selection and Akaike's information criterion (AIC), the general theory and its analytical extension: Psychometrika, v. 52, p. 345–370.

Bredehoeft, J.D., 2003, From models to performance assessment—The conceptual problem: Ground Water, v. 41, no. 5, p. 571–577.

Burnham, K.P., and Anderson, D.R., 2002, Model selection and multi-model inference—A practical information-theoretic approach: New York, Springer-Verlag, 488 p.

Burnham, K.P., and Anderson, D.R., 2004, Multi-model inference—Understanding AIC and BIC model selection: Sociological Methods and Research, v. 33, no. 2, p. 261–304.

Carrera, J., and Neuman, S.P, 1986, Estimation of aquifer parameters under transient and steady state conditions: 1. Maximum likelihood method incorporating prior information: Water Resources Research, v. 22, p. 199–210.

Christensen, S., and Cooley, R.L., 2004, User guide to the UNC process and three utility programs for computation of nonlinear confidence and prediction intervals using MODFLOW-2000: U.S. Geological Survey Techniques and Methods Report, book 6, chap. A10, 186 p.

Constable, S.C., Parker, R.L., and Constable, C.G., 1987, Occam's Inversion—A practical algorithm for generating smooth models from EM sounding data: Geophysics, 52, p. 289–300.

Cooley, R.L., 1983, Incorporation of prior information on parameters into nonlinear regression groundwater flow models, 2, Applications: Water Resources Research, v. 19, no. 3, p. 662–676.

Cooley, R.L., 2004, A theory for modeling ground-water flow in heterogeneous media: U.S. Geological Survey Professional Paper 1679, 220 p.

Cooley, R.L., and Naff, R.L, 1990, Regression modeling of ground-water flow: U.S. Geological Survey Techniques of Water-Resources Investigations, book 3, chap. B4, 232 p.

deLeeuw, J., 1992, Introduction to Akaike (1973) information theory and an extension of the maximum likelihood principle, *in* Kotz, S., and Johnson, N.L., eds., Breakthroughs in statistics: London, Springer-Verlag, v. 1, p. 599–609.

Doherty, J., 2003, Ground water model calibration using pilot points and regularization: Ground Water, v. 41, no. 2, p. 170–177.

Doherty, J., 2004, PEST—Model-Independent Parameter Estimation, User Manual, 5th ed.: Corinda, Australia, Watermark Numerical Computing, 336 p.

Draper, N.R., and Smith, Harry, 1998, Applied regression analysis, 3d ed.: Hoboken, New Jersey, John Wiley and Sons, 706 p.

Fisher, R.A., 1922, On the mathematical foundation of theoretical statistics: Philosophical Transactions of the Royal Society of London, Seris A, v. 222, p. 309–368.

Franssen, H.J., Gomez-Hernandez, Jaime, and Sahuquillo, Andres, 2003, Coupled inverse modeling of groundwater flow and mass transport and the worth of concentration data: Journal of Hydrology, v. 281, p. 281–295.

Gómez-Hernández, J.J., 2006, Complexity: Ground Water, v. 44, no. 6, p. 782–785. doi:10.1111/j.1745-6584.2006.00222.x

Hannan, E.J., and Quinn, B.G., 1979, The determination of the order of an autoregression: Journal of the Royal Statistical Society, Series B 41, p. 190–195.

Harbaugh, A.W., Banta, E.R., Hill, M.C., and McDonald, M.G., 2000, MODFLOW-2000, The U.S. Geological Survey modular ground-water model—Users guide to modularization concepts and the ground-water flow process: U.S. Geological Survey Open-File Report 2000-92, 121 p.

Helsel, D.R., and Hirsch, R.M., 2002, Statistical methods in water resources: U.S. Geological Survey Techniques of Water-Resources Investigations, book 4, chap. A3, available at http://pubs.water.usgs.gov/twri4a3.

Hill, M.C., 1992, A computer program (MODFLOWP) for estimating parameters of a transient, three-dimensional, ground-water flow model using nonlinear regression: U.S. Geological Survey Open-File Report 91–484.

Hill, M.C., 1994, Five computer programs for testing weighted residuals and calculating linear confidence and prediction intervals on results from the ground-water parameter estimation computer program MODFLOWP: U.S. Geological Survey Open-File Report 93–481, 81 p.

Hill, M.C., 1998, Methods and guidelines for effective model calibration: U.S. Geological Survey Water-Resources Investigations Report 98–4005, 90 p.

Hill, M.C., 2006, The practical use of simplicity in developing ground-water models: Ground Water, v. 44, no. 6, p. 775–781.

Hill, M.C., and Tiedeman, C.R., 2007, Effective groundwater model calibration, with analysis of sensitivities, predictions, and uncertainty: New York, Wiley and Sons, 455 p.

Hunt, R.J., Doherty, J., and Tonkin, M.J., 2007, Are models too simple? Arguments for increased parameterization: Ground Water, v. 45, no. 3, p. 254–262. doi:10.1111/j.1745-6584.2007.00316.x

Hurvich, C.M., and Tsai, C-L., 1989, Regression and time series model selection in small samples: Biometrika, v. 76, no. 2, p. 297–307.

Hurvich, C.M., and Tsai, C-L., 1994, Autoregressive model selection in small samples using a bias-corrected version of AIC, in Bozdogan, ed., Engineering and Scientific Applications, v. 3, p. 17–157 H—Proceedings of the First US/Japan Conference on the Frontiers of Statistical Modeling, An Informational Approach: Dordrecht, The Netherlands, Kluwer Academic Publishers.

Kashyap, R.L., 1982, Optimal choice of AR and MA parts in autoregressive moving average models: IEEE Transactions on Pattern Analysis and Machine Intelligence, v. 4, p. 99–104.

Kitanidis, P.K., 1997, Introduction to geostatistics, Applications in hydrogeology: Cambridge University Press, 249 p.

Kullback, S., and Leibler, R.A., 1951, On information and sufficiency: Annals of Mathematical Statistics, v. 22, p. 79–86.

Linhart, H., and Zucchini, W., 1986, Model selection: New York, John Wiley & Sons, 301 p.

Matott, L.S., 2005, OSTRICH, An optimization software tool, documentation and user's guide, Version 1.6: State University of New York at Buffalo, 114 p. Accessed December 28, 2005, at http://www.groundwater.buffalo.edu/software/Ostrich/OstrichMain.html.

McQuarrie, A.D.R., and Tsai, C-L., 1998, Regression and time series model selection: Singapore, World Scientific Publishing Company, 480 p.

Minsker, B., editor, 2003, Long-term ground-water monitoring, The state of the art: American Society of Civil Engineers, stock number 40678. Available at http://www.pubs.asce.org/BOOKdisplay.cgi?9991614.

Moore, C., and Doherty, J., 2005, Role of the calibration process in reducing model predictive error: Water Resources Research, v. 41, DOI: 10.1029/2004WR003501.

Moore, C., and Doherty, J., 2006, The cost of uniqueness in groundwater model calibration: Advances in Water Resources, v. 29, no. 4, p.:605–623.

Neuman, S.P., 2003, Maximum likelihood Bayesian averaging of uncertain model predictions: Stochastic Environmental Research and Risk Assessment, v. 17, p. 291–305.

Neuman, S.P., and Wierenga, P.J., 2003, A comprehensive strategy of hydrogeologic modeling and uncertainty analysis for nuclear facilities and sites: U.S. Nuclear Regulatory Commission NUREG/CR-6805, 236 p.

Poeter, E.P., and Anderson, D.R., 2005, Multi-model ranking and inference in ground-water modeling: Ground Water, v. 43, no. 4, p. 597–605.

Poeter, E.P., and Hill, M.C., 1996, Unrealistic parameter values in inverse modeling, A problem or benefit for model calibration, in Calibration and reliability in groundwater modeling, Proceedings of the 1996 Model CARE Conference, Golden, Colorado, September 1996: IAHS Publication 237, p. 277–285.

Poeter, E.P., and Hill, M.C., 1997, Inverse methods—A necessary next step in ground water modeling: Ground Water, v. 35, no. 2, p. 250–260.

Poeter, E.P., Hill, M.C., Banta, E.R., Mehl, Steffen, and Christensen, Steen, 2005, UCODE_2005 and six other computer codes for universal sensitivity analysis, calibration, and uncertainty evaluation: U.S. Geological Survey Techniques and Methods, book 6, chap. A11, 283 p.

Poeter, E.P., and McKenna, S.A., 1995, Reducing uncertainty associated with groundwater flow and transport predictions: Ground Water 33(6):899–904. Accessed June 4, 2007 at http://www.mines.edu/~epoeter/pubs/1995/gw-reducing.

Sakamoto, Y., Ishiguro, M., and Kitagawa, G., 1986, Akaike Information Criterion Statistics: Tokyo, KTK Scientific Publishers, 320 p.

Saltelli, A., Chan, K., and Scott, E.M., 2000, Sensitivity analysis: New York, Wiley.

Saltelli, A., Tarantola, S., Campolongo, F., and Ratto, M., 2004, Sensitivity analysis in practice: New York, John Wiley and Sons, 232 p.

Schwarz, G., 1978, Estimating the dimension of a model: Annals of Statistics, v. 6, p. 461–464.

Stone, M., 1977, An asymptotic equivalence of choice of model by cross-validation and Akaike's criterion: Journal of the Royal Statistical Society, Series B 39, p. 44–47.

Sugiura, N., 1978, Further analysis of the data by Akaike's information criterion and the finite corrections: Communications in Statistics, Theory and Methods, v. A7, p. 13–26.

Tarantola, Albert, 2005, Inverse problem theory and methods for model parameter estimation: Philadelphia, Society for Industrial and Applied Mathematics, 342 p.

Tiedeman, C.R., Ely, D.M., Hill, M.C., and O'Brien, G.M., 2004, A method for evaluating the importance of system state observations to model predictions, with application to the Death Valley regional groundwater flow system: Water Resources Research, v. 40, w12411, doi:10.1029/2004wr003313.

Tonkin, M.J., and Doherty, John, 2005, A hybrid regularized inversion methodology for highly parameterized environmental models: Water Resources Research, v. 41, W10412, doi:10.1029/2005WR003995.

Tonkin, Matthew J., Tiedeman, Claire R., Ely, D. Matthew, and Hill, Mary C., in press, OPR-PPR, a computer program for assessing data importance to model predictions using linear statistics: U.S. Geological Survey Techniques and Methods, book 6, chapter E2.

Valstar, J.R., McLaughlin, D., and Stroet, C.B.M., 2004, A representer-based inverse method for groundwater flow and transport applications: Water Resources Research, v. 40, no. 5, W05116. DOI:10.1029/2002WR002922.

Wagner, B.J., and Harvey, J.W., 1997, Experimental design for estimating parameters of rate-limited mass transfer, Analysis of stream tracer studies: Water Resources Research, v. 33, no. 7, p. 1731–1742.

Ye, Ming, Neuman, S.P., and Meyer, P.D., 2004, Maximum likelihood Bayesian averaging of spatial variability models in unsaturated fractured tuff: Water Resources Research, v. 40, W05113doi:10.1029/2003WR002557.

Ye, Ming, Neuman, S.P., Meyer, P.D., and Pohlmann, K., 2005, Sensitivity analysis and assessment of prior model probabilities in MLBMA with application to unsaturated fractured tuff: Water Resources Research, v. 41, W12429doi:10.1029/2005WR004260.

Yeh, WW-G., and Yoon, Y.S., 1981, Aquifer parameter identification with optimum dimension in parameterization: Water Resources Research, v. 7, no. 3, p. 664–672.

Appendix 1: Connection of MMA with the JUPITER API

The JUPITER API is a computer programming environment that includes conventions and software components designed to support the development of computer programs that perform model sensitivity analysis, data needs evaluation, sensitivity analysis, uncertainty evaluation, and(or) optimization. The goal of the JUPITER API is to allow scientists to be able to express their ideas in useful programs that can be readily applied to practical problems. It is hoped that this facilitation of the connection between research and application will accelerate technical advance of using data to model natural systems and improve the scientific basis, and, therefore, the success, of societal decisions about these systems.

MMA uses the following JUPITER modules indicated in Table A1-1 and described in chapters of Banta and others (2006):

Table A1-1. JUPITER API modules used by MMA and the related chapter and chapter authorship in Banta and others (2006).

Modules	Chapter	Authorship of chapter
Datatypes	3	Banta
Global Data	4	Banta and Doherty
Utilities	5	Banta, Doherty, and Poeter
Basic	6	Banta and Doherty
Equation	9	Doherty
Data-Exchange Files	17	Poeter, Banta, and Hill

In addition, because of dependencies in the UCODE_2005 modules used by MMA, four additional JUPITER modules need to be included when compiling MMA. They are the Prior-Information, Dependents, Statistics, and Sensitivity modules. Thus, except for the Parallel-Processing and Model Input and Output modules, all of the original JUPITER modules need to be included when compiling MMA.

Appendix 2: Example Problem

The system for the example problem is discussed by Poeter and Anderson (2005); a subset of the models they discuss are presented here. The system is two-dimensional, unconfined, and steady state. The model domain is 5,000 meters (m) in the east-west direction and 3,000 m north and south, as shown in Figure A2-1. This example is synthetic; there is one "true" model and many calibrated models. All are simulated with MODFLOW-2000 (Harbaugh and others, 2000). The following simulated boundary conditions and stresses are imposed in all models:

- A no-flow boundary is defined on the northern, western, and southern boundaries.
- The aquifer base is at –10 m.
- There is a 10-m-wide tributary in the center of the watershed. Its stage ranges from 20 m at the west end, decreasing linearly to 5 m where the tributary intersects the river. The streambed is 5 m thick and has a vertical hydraulic conductivity of 0.2 meter per day (m/d).
- A 10-m-wide river bounds the east edge of the simulated system. The stage is 5 m, the streambed is 5 m thick, and the bottom of the sediment layer is at an elevation of 0 m. The vertical hydraulic conductivity is 0.1 m/d
- Rivers are simulated as head-dependent flux boundaries using the MODFLOW-2000 River Package.
- Recharge is applied uniformly to the entire top of all models.
- One well pumps 2,000 m^3/d for calibration and prediction conditions. For predictive conditions, a second well pumps 3,000 m^3/d. See Figure A2-1 for the well locations.

The synthetic "true" model has a grid of 250 × 150 cells, each 20m × 20m. The uniform recharge rate is 8×10^{-4} m/d. There are five zones of hydraulic conductivity (K), as shown in Figure A2-1 and A2-2. The true head distributions for calibration and predictions conditions are shown in Figure A2-1 and A2-2, respectively.

All models used for calibration and prediction have grids that consist of 50 × 30 cells, each 100m × 100m, as shown in Figure A2-3.

The "observed" heads and flow are derived from the detailed "true" synthetic model. There are 20 head observations located as shown in Figure A2-3. The one flow observation represents streamflow gains along the tributary shown in Figure A2-3. Table A2-1 relates the observation names to the sequential numbers shown in Figure A2-3.

Each model has three, four, or five estimated parameters. The uniform recharge rate is estimated in each of the calibrated models. Two, three, or four hydraulic-conductivity parameters are estimated in each model. Each hydraulic-conductivity value applies to a zone of constant value composed of grid cells that generally are not adjacent to one another. For each number of hydraulic conductivity parameters, five different zonal distributions are considered, each of which is obtained from indicator kriging. The zones are defined in files named za.dat, zb.dat, zc.dat, and so on. There are 15 calibrated models named using the number of zones (2, 3, or 4) followed by the letter A, B, C, D, or E. Example model names are 2A and 4E.

All 15 models are used to estimate parameters with observations only, and are analyzed as a set using MMA.

Appendix 2: Example Problem

Five of the models also are used to estimate parameters with observations and prior information equations on the parameters, and these also are analyzed as a set using MMA.

The five models calibrated with both observations and prior information on the parameters are the models with two hydraulic-conductivity parameters. Except for the addition of the prior information, they are identical to the five models with two hydraulic-conductivity parameters calculated with observations only. The low K zone (K1) is assigned prior information of 4 m/d and the high K zone (K5) is assigned 22 m/d. The statistics used to weight the prior information are variances of 0.25, which for these log-transformed parameters are consistent with a 95-percent confidence interval equivalent to plus and minus one order of magnitude.

The 22 predictions are 20 heads each located 200 m west of one of the 20 head-observation locations and 2 flow predictions of net ground-water discharge, one to the river and the other to the tributary. The predictions are affected by additional pumping of 3,000 cubic meters per day (m^3/d), located as shown in Figure A2-3. The head prediction names are formed by adding "off" to the end of the name of the observation 200 m to the east (Table A2-1). The flow predictions are names qmain for the river and qtrib for the tributary.

UCODE_2005 is used to estimate parameters and simulate predictions. Linear confidence intervals on the predictions are calculated using the computer program LINEAR_UNCERTAINTY, which is distributed with UCODE_2005. Both programs are documented by Poeter and others (2005). All the individual-model regression, prediction, prediction uncertainty runs, and the model-analysis run can be repeated by executing the batch files described below and in Appendix 3.

Three MMA main input files are in the MMA distribution, and two of these contain evaluations the models calibrated only with observations. One of these contains minimal input and uses all defaults in MMA except that VERBOSE=0, and is referred to as the minimal run. The second uses many options provided by MMA, and is referred to as the extensive run. The third MMA input file evaluates the five models that include prior information on parameters to provide an example of MMA results when prior information is included. The MMA main input files are named

"testmma_min", "testmma_ext", and "testmma_pri".

Batch files that run MMA for each input file are named, respectively,

"mma_minimal.bat", "mma_extensive.bat", and "mma_prior.bat".

There is also a batch file that removes all output files from that execution of MMA. The name of this batch file is

"clean_mma.bat".

The file testmma._ext is listed below as an example.

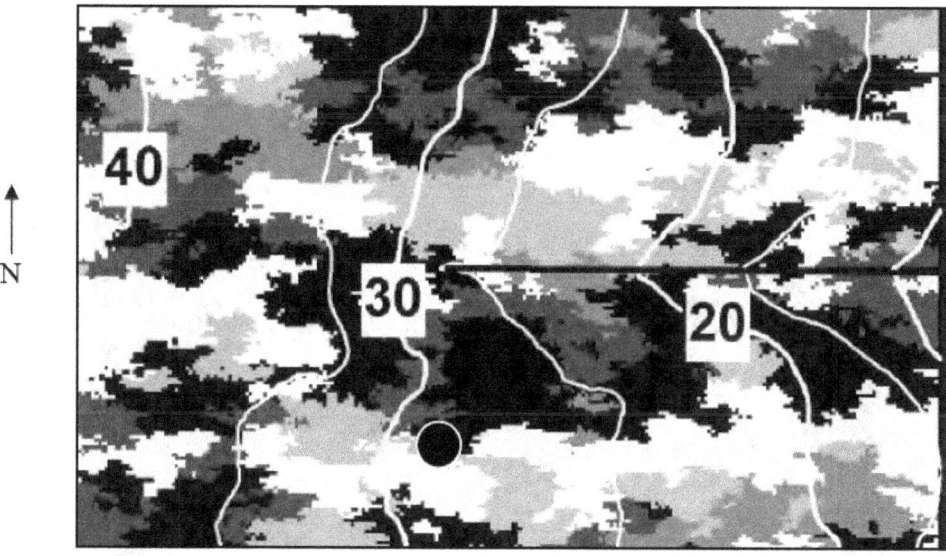

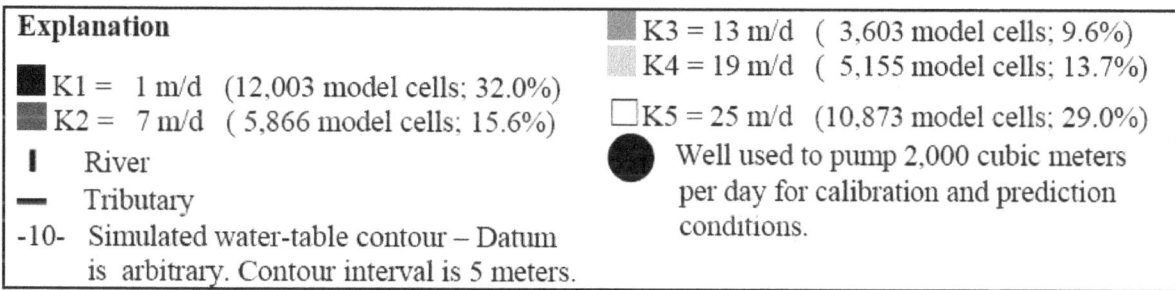

Figure A2-1. True heterogeneity and head distribution for the synthetic model under hydraulic conditions used to generate calibration data. The area is 5,000 meters by 3,000 meters (from Poeter and Anderson, 2005).

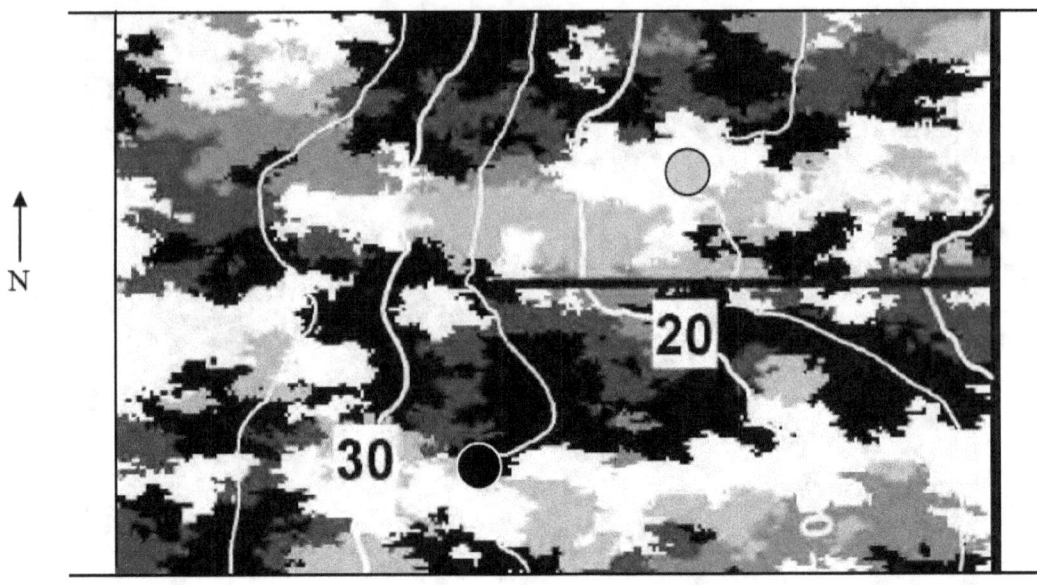

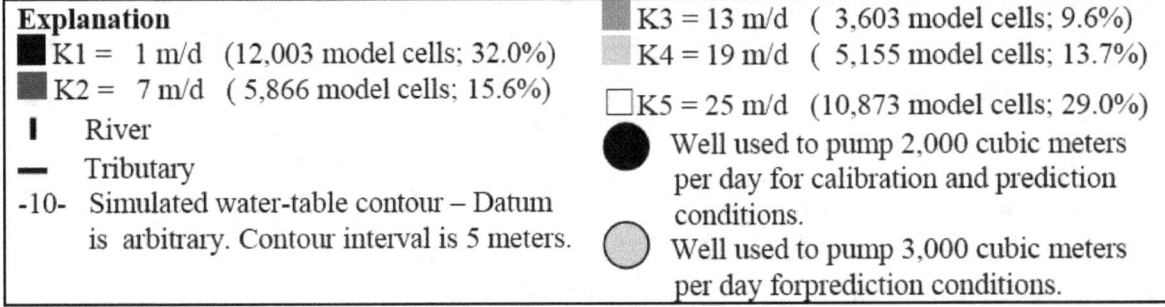

Explanation
- K1 = 1 m/d (12,003 model cells; 32.0%)
- K2 = 7 m/d (5,866 model cells; 15.6%)
- ▮ River
- — Tributary
- -10- Simulated water-table contour – Datum is arbitrary. Contour interval is 5 meters.
- K3 = 13 m/d (3,603 model cells; 9.6%)
- K4 = 19 m/d (5,155 model cells; 13.7%)
- ☐ K5 = 25 m/d (10,873 model cells; 29.0%)
- ● Well used to pump 2,000 cubic meters per day for calibration and prediction conditions.
- ◯ Well used to pump 3,000 cubic meters per day forprediction conditions.

Figure A2-2. True heterogeneity and head distribution for the synthetic model under predictive conditions. Bold, black lines indicate the location of a tributary and a river. The area is 5,000 meters by 3,000 meters (from Poeter and Anderson, 2005).

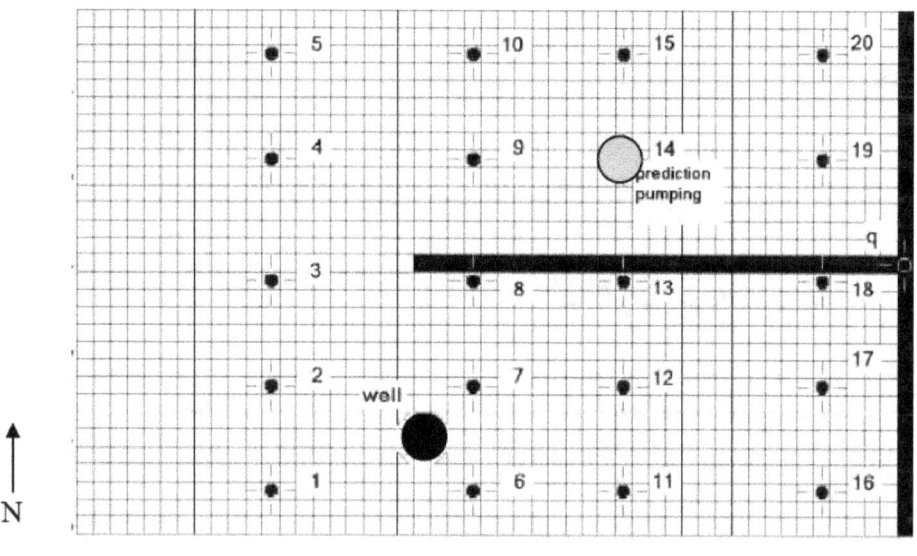

Explanation

❙ River

— Tributary

-10- Simulated water-table contour – Datum
is arbitrary. Contour interval is 5 meters.

● Well used to pump 2,000 cubic meters per day for calibration and prediction conditions.

○ Well used to pump 3,000 cubic meters per day for prediction conditions. Observation 14 is at the same location.

Figure A2-3. Coarse grid used for all calibrated models (from Poeter and Anderson, 2005).

Table A2-1. The sequential location numbers shown in Figure A2-3 and the names used for observations at these locations.

[Prediction names are formed by adding "off" to the end of the associated observation name]

Number	Name	Number	Name	Number	Name	Number	Name
5	o15	10	o25	15	o35	20	o45
4	o14	9	o24	14	o34	19	o44
3	o13	8	o23	13	o33	18	o43
2	o12	7	o22	12	o32	17	o42
1	o11	6	o21	11	o31	16	o41

Input Files

Two main input files and one of the data-exchange files from one of the models are listed. The main input file listed first is testmma_min, which contains only the required input block; the second is testmma_ext, which uses more features of MMA.

The testmma_min Main Input File

```
BEGIN OPTIONS
  Verbose=0
END OPTIONS

BEGIN MODEL_PATHS TABLE
nrow=15  ncol=1 columnlabels
PathAndRoot
..\..\data-win\obsonly\z2\1\z
..\..\data-win\obsonly\z2\2\z
..\..\data-win\obsonly\z2\3\z
..\..\data-win\obsonly\z2\4\z
..\..\data-win\obsonly\z2\5\z
..\..\data-win\obsonly\z3\1\z
..\..\data-win\obsonly\z3\2\z
..\..\data-win\obsonly\z3\3\z
..\..\data-win\obsonly\z3\4\z
..\..\data-win\obsonly\z3\5\z
..\..\data-win\obsonly\z4\1\z
..\..\data-win\obsonly\z4\2\z
..\..\data-win\obsonly\z4\3\z
..\..\data-win\obsonly\z4\4\z
..\..\data-win\obsonly\z4\5\z
END MODEL_PATHS
```

The testmma_ext Main Input File

```
BEGIN OPTIONS
  Verbose=0
END OPTIONS

BEGIN OUTPUT_CONTROL
WritePreds = yes
WRITEPARAMREGRESS=yes
WRITEPARAMNATIVE=yes
END OUTPUT_CONTROL

BEGIN MODEL_GROUPS
GroupName=z2  avg=yes
GroupName=z3  avg=yes
GroupName=z4  avg=yes
END MODEL_GROUPS
```

```
BEGIN PARAM_EQNS TABLE
nrow=6  ncol=3 columnlabels
ParEqnName      ParEqn              GroupName
Crit1_Z2a       Kx5.gt.Kx1          z2
Crit1_Z3a       Kx5.gt.Kx3          z3
Crit1_Z3b       Kx3.gt.Kx1          z3
Crit1_Z4a       Kx5.gt.Kx4          z4
Crit1_Z4b       Kx4.gt.Kx2          z4
Crit1_Z4c       Kx2.gt.Kx1          z4
END PARAM_EQNS

BEGIN PARAM_AVGS TABLE
nrow=12  ncol=3 columnlabels
ParamAvgName  GroupName Avg
Kx1 Z2 yes
Kx5 Z2 yes
RCH Z2 yes
Kx1 Z3 yes
Kx3 Z3 yes
Kx5 Z3 yes
RCH Z3 yes
Kx1 Z4 yes
Kx2 Z4 yes
Kx4 Z4 yes
Kx5 Z4 yes
RCH Z4 yes
END PARAM_AVGS

BEGIN MODEL_PATHS TABLE
nrow=15  ncol=2 columnlabels
PathAndRoot
..\..\data-win\obsonly\z2\1\z     z2
..\..\data-win\obsonly\z2\2\z     z2
..\..\data-win\obsonly\z2\3\z     z2
..\..\data-win\obsonly\z2\4\z     z2
..\..\data-win\obsonly\z2\5\z     z2
..\..\data-win\obsonly\z3\1\z     z3
..\..\data-win\obsonly\z3\2\z     z3
..\..\data-win\obsonly\z3\3\z     z3
..\..\data-win\obsonly\z3\4\z     z3
..\..\data-win\obsonly\z3\5\z     z3
..\..\data-win\obsonly\z4\1\z     z4
..\..\data-win\obsonly\z4\2\z     z4
..\..\data-win\obsonly\z4\3\z     z4
..\..\data-win\obsonly\z4\4\z     z4
..\..\data-win\obsonly\z4\5\z     z4
END MODEL_PATHS

BEGIN PREDS TABLE
nrow=22  ncol=1 columnlabels
Prediction
o11off
o12off
o13off
o14off
o15off
o21off
o22off
o23off
o24off
o25off
o31off
o32off
o33off
o34off
o35off
o41off
```

93

Appendix 2: Example Problem

```
o42off
o43off
o44off
o45off
qtrib
qmain
END PREDS

BEGIN ANALYSES TABLE
nrow=3  ncol=3 columnlabels
AnalysisLabel CritEqn PrEqn
   AICcObs        AICcObs    exp(-0.5*(valcrit-mincrit))
   KICObs         KICObs     exp(-0.5*(valcrit-mincrit))
   SWSRlin        SWSRObs    1.+((mincrit-valcrit)/(maxcrit-mincrit))
END ANALYSES
```

A *Data-Exchange* File

This data-exchange file z._os is from the model with PathAndRoot= ..\..\data-win\obsonly\z2\1\z in the Model_Paths input block.

"SIMULATED EQUIVALENT"	"OBSERVED or PRIOR VALUE"	"PLOT SYMBOL"	"OBSERVATION or PRIOR NAME"
34.03728	33.13374	1	o11
34.49965	34.69764	1	o12
35.94202	37.75531	1	o13
37.76408	35.94014	1	o14
38.95450	39.17625	1	o15
26.93647	25.71865	1	o21
26.99618	23.09924	1	o22
25.48750	24.81570	1	o23
26.68219	25.41486	1	o24
30.88262	29.82160	1	o25
24.77223	21.28041	1	o31
23.70066	24.25597	1	o32
20.89334	20.43605	1	o33
21.76271	21.90114	1	o34
24.28345	24.36283	1	o35
17.81798	13.21173	1	o41
16.32330	17.20430	1	o42
12.45327	12.10146	1	o43
14.90440	15.32061	1	o44
15.53301	15.51246	1	o45
-6991.581	-5568.000	2	qtrib

Output Files

MMA has two types of output files. One output file is designed primarily to be looked at by the user. The filename extension of this file is "#mout." Other output files are primarily intended to be used by other computer programs, including spreadsheets and plotting programs. These files are constructed using the conventions of JUPITER API data-exchange files. Their filename extensions begin with "_". For all of the output files, the first part of the filename is defined in the MMA command line, as explained in chapter 4.

The Output File Designed Primarily to be Looked at by the User

The output file designed primarily to be looked at by the user is also called the MMA main output file in this work. For the more extensive of the examples distributed with MMA, it is named Testmma_ext.#mout.

Testmma_ext.#mout

This is an example of the MMA main output file.

```
--------------------------------------------------------------------------------
--------------------------------------------------------------------------------
Executing MMA,   Version:   1.000

 NAME_OF_ANALYSIS: "testmma_ext"
 DATE_that_ANALYSIS_was_EXECUTED   5- 5-2007 TIME 13: 6:32 ZONE -0600


  All Model Measures
     will be written to  unit:    12
     File: testmma_ext._mma

  All Model Measures
     will be written to  unit:    13
     File: testmma_ext._mma_gstats

  All Model Measures
     will be written to  unit:    14
     File: testmma_ext._mma_xyzt

  Model Ranks for ALL Measures
     will be written to unit:    15
      File: testmma_ext._rank

  Model Ranks for ALL Measures
     will be written to unit:    16
      File: testmma_ext._rank_gstats

  Model Ranks for ALL Measures
     will be written to unit:    17
      File: testmma_ext._rank_xyzt

  ****************************************************************

  ********* SUM OF PRIOR MODEL PROBABILITIES IS   1.00 *********

    SUM =   1.0000000E+00
```

```
FOR MODELS WITH NO USER-ASSIGNED PROBABILITY, THE PROBABILITY
WAS SET TO 1/(NUMBER OF MODELS) AND USED IN THE SUM
***********************************************************

    1 ANALYZED:                            2A in ..\..\data-win\obsonly\z2\1\z
    2 ANALYZED:                            2B in ..\..\data-win\obsonly\z2\2\z
    3 ANALYZED:                            2C in ..\..\data-win\obsonly\z2\3\z
    4 ANALYZED:                            2D in ..\..\data-win\obsonly\z2\4\z
    5 ANALYZED:                            2E in ..\..\data-win\obsonly\z2\5\z
    6 ANALYZED:                            3A in ..\..\data-win\obsonly\z3\1\z
    7    UNREASONABLE PARAMETERS:          3B in ..\..\data-win\obsonly\z3\2\z
    8 ANALYZED:                            3C in ..\..\data-win\obsonly\z3\3\z
    9 ANALYZED:                            3D in ..\..\data-win\obsonly\z3\4\z
   10 ANALYZED:                            3E in ..\..\data-win\obsonly\z3\5\z
   11    UNREASONABLE PARAMETERS:          4A in ..\..\data-win\obsonly\z4\1\z
   12    UNREASONABLE PARAMETERS:          4B in ..\..\data-win\obsonly\z4\2\z
   13                NOT CONVERGED:         4C in ..\..\data-win\obsonly\z4\3\z
   14 ANALYZED:                            4D in ..\..\data-win\obsonly\z4\4\z
   15    UNREASONABLE PARAMETERS:          4E in ..\..\data-win\obsonly\z4\5\z

     15 MODELS were evaluated, of those:
         14 models CONVERGED and
          1 models DID NOT CONVERGE


     14 MODELS CONVERGED, of those:
         10 models had REASONABLE PARAMETERS and
          4 models had UNREASONABLE PARAMETERS


     10 MODELS will be ranked and weighted
```



```
    CHECK THAT PARAMETERS TO BE AVERAGED EXIST FOR ALL MODELS IN THE GROUP


  NAMES OF ANALYZED MODELS   NORMALIZED PRIOR MODEL PROBABILITY
  ------------------------   ----------------------------------
                       2A                0.1000
                       2B                0.1000
                       2C                0.1000
                       2D                0.1000
                       2E                0.1000
                       3A                0.1000
                       3C                0.1000
                       3D                0.1000
                       3E                0.1000
                       4D                0.1000
```


```
Analysis: AICcObs      will be written to:  "testmma_ext._anals_AICcObs"

 Model-averaged Predictions
   for Analysis: AICcObs
   will be written to unit:     17
   File: testmma_ext._preds_AICcObs


       Predictions involved in averaging
          will be written to unit:     17
          File: testmma_ext._IndividPreds
```

```
           Prediction Variances involved in averaging
              will be written to unit:      18
              File: testmma_ext._IndividPredsVar


   Model-averaged Parameters
     for Analysis: AICcObs
     will be written to unit:      17
     File: testmma_ext._params_AICcObs



         NATIVE Parameter values involved in averaging
            will be written to unit:      18
            File: testmma_ext._IndividParamNative

         NATIVE Parameter variances involved in averaging
            will be written to unit:      19
            File: testmma_ext._IndividParVarNative

         REGRESSION SPACE Parameter values involved in averaging
            will be written to unit:      20
            File: testmma_ext._IndividParamRegress

         REGRESSION SPACE variances involved in averaging
            will be written to unit:      21
            File: testmma_ext._IndividParVarRegress

-----------------------------------------------------------------------------
-----------------------------------------------------------------------------

 Analysis: KICObs       will be written to:   "testmma_ext._anals_KICObs"

  Model-averaged Predictions
     for Analysis: KICObs
     will be written to unit:      17
     File: testmma_ext._preds_KICObs


         Predictions involved in averaging
            will be written to unit:      17
            File: testmma_ext._IndividPreds

         Prediction Variances involved in averaging
            will be written to unit:      18
            File: testmma_ext._IndividPredsVar

  Model-averaged Parameters
     for Analysis: KICObs
     will be written to unit:      17
     File: testmma_ext._params_KICObs


     -----------------------------------------------------------------------
     -----------------------------------------------------------------------

 Analysis: SWSRlin      will be written to:   "testmma_ext._anals_SWSRlin"

  Model-averaged Predictions
     for Analysis: SWSRlin
     will be written to unit:      17
     File: testmma_ext._preds_SWSRlin


         Predictions involved in averaging
            will be written to unit:      17
            File: testmma_ext._IndividPreds
```

```
    Prediction Variances involved in averaging
        will be written to unit:    18
        File: testmma_ext._IndividPredsVar

  Model-averaged Parameters
    for Analysis: SWSRlin
    will be written to unit:    17
    File: testmma_ext._params_SWSRlin

-------------------------------------------------------------------------------

  A list of Model Names and Associated Paths for Analyzed Models
    will be written to unit:    17
      File: testmma_ext._ModelNamesPaths

-------------------------------------------------------------------------------
  MMA COMPLETED SUCCESSFULLY
-------------------------------------------------------------------------------
```

MMA Data-Exchange Output Files

There are up to 15 data-exchange files produced by MMA. Here, selections from the nine produced by the "extensive" run are shown. This run uses many of the features of MMA.

Testmma_ext._mma

This file contains the variables and measures of overall model fit described in chapter 3 and listed in Table 8.

A data set with the following header is always included in this file.

```
"ID#"
"MODEL         " "NPE        " "NOBS       " "NPR       " "SWSROBS       "
"CEVOBS        " "MLOFOBS    " "AICOBS     " "AICCOBS   " "BICOBS        "
"KICOBS        " "XTWXOBS    " "PATHANDROOT "
```

If prior information equations are defined and the prior information equations are the same for all models so that MMA considers prior information, three additional data sets are printed in this file. They have the following headers.

```
"ID#"
"MODEL         " "NPE        " "NOBS       " "NPR       " "SWSRWPRI      "
"CEVWPRI       " "MLOFWPRI   " "AICWPRI    " "AICCWPRI  " "BICWPRI       "
"KICWPRI       " "XTWXWPRI   " "PATHANDROOT "

"ID#"
"MODEL         " "NPE        " "NOBS       " "NPR       " "SWSR_PR-O     "
"CEV_PR-O      " "MLOF_PR-O  " "AIC_PR-O   " "AICC_PR-O " "BIC_PR-O      "
"KIC_PR-O      " "XTWX_PR-O  " "PATHANDROOT "

"ID#"
"MODEL         " "NPE        " "NOBS       " "NPR       " "SWSR_%        "
"CEV_%         " "MLOF_%     " "AIC_%      " "AICC_%    " "BIC_%         "
"KIC_%         " "XTWX_%     " "PATHANDROOT "
```

Testmma_ext._mma_gstats

A data set with the following header is always included in this file.

```
"ID#"
"MODEL         " "R2_OSOBS    " "INT_OSOBS   " "SLP_OSOBS   " "R2_WSOBS     "
"INT_WSOBS     " "SLP_WSOBS   " "R2_WWOBS    " "INT_WWOBS   " "SLP_WWOBS    "
"R2_NMOBS      " "PATHANDROOT "
```

If prior information equations are defined and the prior information equations are the same for all models so that MMA considers prior information, three additional data sets are printed in this file. They have the following headers.

```
"ID#"
"MODEL          " "R2_OSWPRI   " "INT_OSWPRI  " "SLP_OSWPRI  " "R2_WSWPRI    "
"INT_WSWPRI   " "SLP_WSWPRI  " "R2_WWWPRI   " "INT_WWWPRI  " "SLP_WWWPRI   "
"R2_NMWPRI    " "PATHANDROOT "

"ID#"
"MODEL          " "R2_OSPR-O   " "INT_OSPR-O  " "SLP_OSPR-O  " "R2_WSPR-O    "
"INT_WSPR-O   " "SLP_WSPR-O  " "R2_WRWOPR-O " "INT_WRWOPR-O" "SLP_WRWOPR-O"
"RN2PR-O      " "PATHANDROOT "

"ID#"
"MODEL          " "R2_OS_%     " "INT_OS_%    " "SLP_OS_%    " "R2_WS_%      "
"INT_WS_%     " "SLP_WS_%    " "R2_WRWO_%   " "INT_WRWO_%  " "SLP_WRWO_%   "
"RN2_%        " "PATHANDROOT "
```

Testmma_ext._mma_xyzt

A data set with the following header is always included in this file.

```
"ID#"
"MODEL          " "R2_RT       " "INT_RT      " "SLP_RT      " "R2_WRT       "
"INT_WRT      " "SLP_WRT     " "CNT_LOCX    " "CNT_LOCY    " "CNT_LOCZ     "
"DIFCNT_WX    " "DIFCNT_WY   " "DIFCNT_WZ   " "DISTCNTW    " "DIFCNT_SGNWX"
"DIFCNT_SGNWY" "DIFCNT_SGNWZ" "DISTCNT_SGNW" "DIFCNT_MGWX " "DIFCNT_MGWY "
"DIFCNT_MGWZ " "DIST_CNTMGW " "PATHANDROOT "
```

Testmma_ext._rank

A data set with the following header is always included in this file.

```
"ID#"
"MODEL          " "NPE         " "NOBS        " "NPR         " "SWSROBS      "
"CEVOBS       " "MLOFOBS     " "AICOBS      " "AICCOBS     " "BICOBS       "
"KICOBS       " "XTWXOBS     " "PATHANDROOT "
```

If prior information equations are defined and the prior information equations are the same for all models so that MMA considers prior information, two additional data sets are printed in this file. They have the following headers.

```
"ID#"
"MODEL          " "NPE         " "NOBS        " "NPR         " "SWSRWPRI     "
"CEVWPRI      " "MLOFWPRI    " "AICWPRI     " "AICCWPRI    " "BICWPRI      "
"KICWPRI      " "XTWXWPRI    " "PATHANDROOT "

"ID#"
"MODEL          " "NPE         " "NOBS        " "NPR         " "SWSR_CHG     "
"CEV_CHG      " "MLOF_CHG    " "AIC_CHG     " "AICC_CHG    " "BIC_CHG      "
"KIC_CHG      " "XTWX_CHG    " "PATHANDROOT "
```

Testmma_ext._rank_gstats

A data set with the following header is always included in this file.

```
"ID#"
"MODEL         " "R2_OSOBS   " "INT_OSOBS   " "SLP_OSOBS   " "R2_WSOBS     "
"INT_WSOBS   " "SLP_WSOBS   " "R2_WWOBS   " "INT_WWOBS   " "SLP_WWOBS   "
"R2_NMOBS   " "PATHANDROOT "
```

If prior information equations are defined and the prior information equations are the same for all models so that MMA considers prior information, two additional data sets are printed in this file. They have the following headers.

```
"ID#"
"MODEL         " "R2_OSWPRI   " "INT_OSWPRI   " "SLP_OSWPRI   " "R2_WSWPRI     "
"INT_WSWPRI   " "SLP_WSWPRI   " "R2_WWWPRI   " "INT_WWWPRI   " "SLP_WWWPRI   "
"R2_NMWPRI   " "PATHANDROOT "

"ID#"
"MODEL         " "R2_OS_CHG   " "INT_OS_CHG   " "SLP_OS_CHG   " "R2_WS_CHG     "
"INT_WS_CHG   " "SLP_WS_CHG   " "R2_WW_CHG   " "INT_WW_CHG   " "SLP_WW_CHG   "
"R2_NM_CHG   " "PATHANDROOT "
```

Testmma._rank_xyzt

A data set with the following header is always included in this file.

```
"ID#"
"MODEL         " "R2_RT       " "INT_RT      " "SLP_RT      " "R2_WRT       "
"INT_WRT     " "SLP_WRT     " "CNT_LOCX    " "CNT_LOCY    " "CNT_LOCZ     "
"DIFCNT_WX   " "DIFCNT_WY   " "DIFCNT_WZ   " "DISTCNTW    " "DIFCNT_SGNWX"
"DIFCNT_SGNWY" "DIFCNT_SGNWZ" "DISTCNT_SGNW" "DIFCNT_MGWX " "DIFCNT_MGWY "
"DIFCNT_MGWZ " "DIST_CNTMGW " "PATHANDROOT "
```

Testmma._anals_AICcObs

A data set with the following header is always included in this file. The AICcObs in the labels in the file and in the filename may be replaced by other analysis names. If default analysis types are used, and there are no prior information equations in the evaluated models (or they do not all have the same prior) then the file for AICcObs is printed with two header lines as shown below.

```
"ANALYSIS NAME:" "AICcObs" "Criterion Equation:" "aiccobs" "Weighting Equation:"
"exp(-0.5*(valcrit-mincrit))"

"MODEL"         "PRIOR PROB"     "CRITERION" "RANK"    "PROBABILTY"       "DELTA"
"EVIDENCE-RATIO" "ER-INVERSE as %" "PATHANDROOT"
```

Testmma_ext_params_AICcObs

To fit the file onto this page, two changes are made to the column labels in the second and similar lines: averaged is replaced by avg; confidence is replaced by conf. The number of spaces in the following lines of data also is altered. The entire file is shown.

```
"Number of models: " "  5"    "AICcObs MODEL-AVG PARAMETERS"
"PARAMETER " "Model-Avg Lower Conf" "Model-Avg Value" "Model-Avg Upper Conf" "Model-Avg Variance" "ESTIMATION STATE"
"GROUP: Z2"
Kx1     4.114505        6.264521        9.538020        1.850562        TRANSFORMED
Kx5    28.33907        37.60670        49.90508        29.13423        TRANSFORMED
RCH     7.8837356E-04   8.7065446E-04   9.5293536E-04   1.6925367E-09   NATIVE

"Number of models: " "  4"    "AICcObs MODEL-AVG PARAMETERS"
"PARAMETER " "Model-Avg Lower Conf" "Model-Avg Value" "Model-Avg Upper Conf" "Model-Avg Variance" "ESTIMATION STATE"
"GROUP: Z3"
Kx1     1.768443        3.077032        5.353932        0.8138333       TRANSFORMED
Kx3    12.81338        19.85061        30.75277        20.26080        TRANSFORMED
Kx5    27.18498        47.31175        82.33965        192.5789        TRANSFORMED
RCH     7.4476371E-04   8.0660791E-04   8.6845211E-04   9.5617619E-10   NATIVE

"Number of models: " "  1"    "AICcObs MODEL-AVG PARAMETERS"
"PARAMETER " "Model-Avg Lower Conf" "Model-Avg Value" "Model-Avg Upper Conf" "Model-Avg Variance" "ESTIMATION STATE"
"GROUP: Z1"
Kx1     1.723412        3.088962        5.536509        0.9219889       TRANSFORMED
Kx2     7.831663        17.60116        39.55745        64.92938        TRANSFORMED
Kx4    12.57224        21.41289        36.47017        36.11301        TRANSFORMED
Kx5    25.80325        45.98509        81.95203        199.8509        TRANSFORMED
RCH     7.4440904E-04   8.0734190E-04   8.7027476E-04   9.9013613E-10   NATIVE
```

Testmma_ext_preds_AICcObs

To fit the file onto this page, two changes are made to the column labels in the second line: "predicted" and "predictions" are replaced by "preds"; and "confidence" is replaced by "conf". The number of spaces in the following lines of data also is altered. Only three of the many lines of data are shown. The dots represent omitted lines.

```
"AICc MODEL-AVERAGED, PREDICTIONS and INDIVIDUAL CONFIDENCE INTERVALS" "Number of models: " " 10"
"PRED NAME" "MOD-AVG PRED VALUE" "MOD-AVG LOWER CONF INT" "MOD-AVG UPPER CCNF INT" "MOD-AVG STANDARD DEVIATION" "PLOT SYMBOL"
o11off  32.19502        30.51133        33.87871        0.8418460       1
o12off  32.97159        31.80821        34.13497        0.5816892       1
o13off  34.55021        33.85958        35.24083        0.3453135       1
.
.
.
"AICc MODEL-AVERAGED, PREDICTIONS and SIMULTANEOUS CONFIDENCE INTERVALS, LIMITED" "Number of models: " " 10"
.
.
.
"AICc MODEL-AVERAGED, PREDICTIONS and SIMULTANEOUS CONFIDENCE INTERVALS, INFINITE" "Number of models: " " 10"
```

Appendix 2: Example Problem

.... "AICc MODEL-AVERAGED, PREDICTIONS and INDIVIDUAL PREDICTION INTERVALS" "Number of models: " " 10"

. . . . "AICc MODEL-AVERAGED, PREDICTIONS and SIMULTANEOUS PREDICTION INTERVALS, LIMITED" "Number of models: " " "   10"

. . . . . . "AICc MODEL-AVERAGED, PREDICTIONS and SIMULTANEOUS PREDICTION INTERVALS, INFINITE" "Number of models: " "   10"

Results

Selected results from output files produced for the extensive MMA run using observations only are shown in this section. The files from which the plotted data are obtained are listed. The files are from the directory test-win\01-Run_MMA, unless otherwise noted.

The model-averaged predictions are compared to the predictions from the six most probable models in Figure A2-4. Model probabilities are presented later in this section. As illustrated, the prediction of the most probable conceptual model may differ substantially from other models with high probabilities, which likely would have been acceptable models if they were the only conceptual model under consideration.

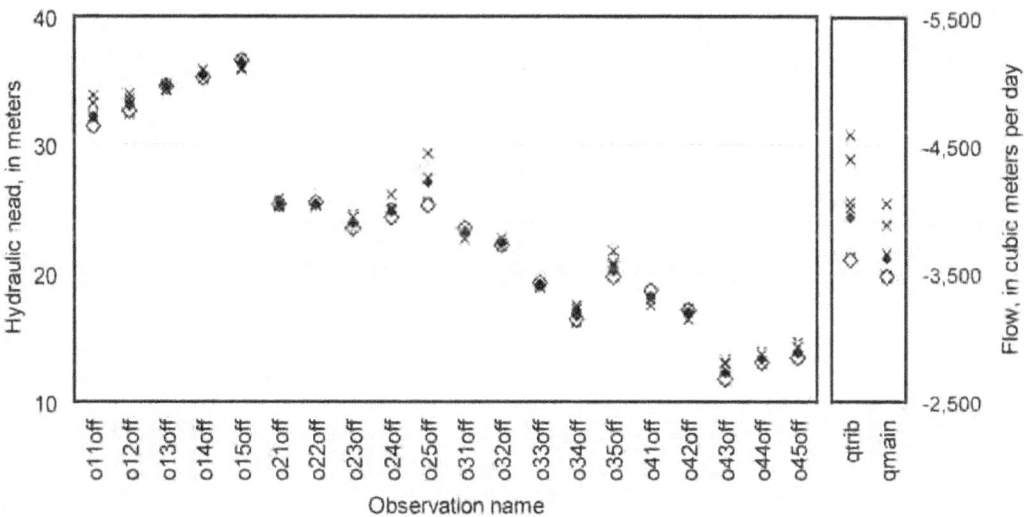

EXPLANATION

◆ Model averaged prediction
◇ Prediction from most probable mode

× Predictions from the next five most probable models, models 2A, 2E, 2C, 2B, and 4D

Figure A2-4. Predictions calculated by model averaging, by model 3D, which is the most probable model, and by the next five most probable models. Negative flows indicate flow out of the ground-water system and into the river.

Model averaged predictions and linear confidence intervals are compared with analogous results from the most probable model in Figure A2-5. Individual and simultaneous intervals are shown. The simultaneous intervals are calculated as the smaller of Sheffé d=k or Bonferroni intervals, as described in Poeter and others (2005, p. 184). Simultaneous intervals tend to be wider than individual intervals, as shown in this example. Here, the model-averaged simultaneous intervals (Figure A2-5A) are between about 5 and 50 percent wider than the model-averaged individual intervals. Though the difference for flows looks large, this is primarily due to the scale used in the graph. The model-averaged simultaneous intervals for flows are both about 25 percent larger than the individual intervals.

(A) Model averaged

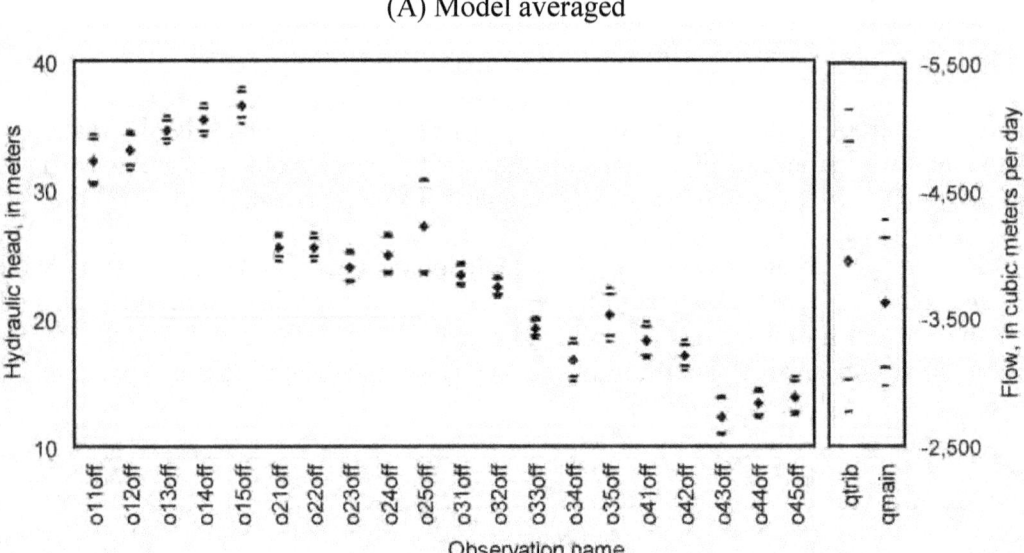

(B) For model 3D, the most probable model

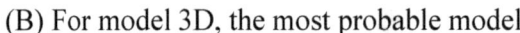

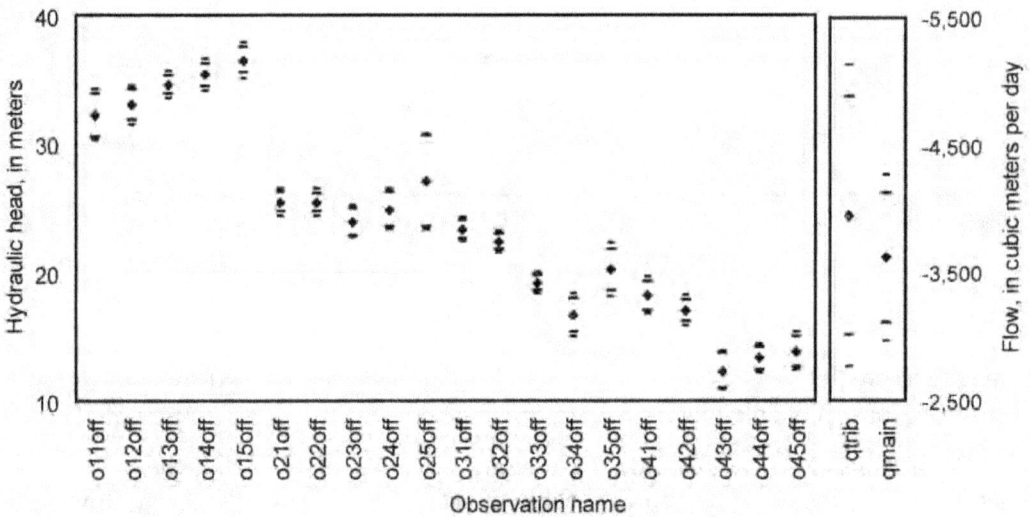

EXPLANATION
◆ Prediction   – Limits of individual confidence intervals
- Limits of simultaneous confidence intervals

Figure A2-5. Predicted heads and 95-percent confidence intervals for (A) model-averaged values from the MMA output file testmma_ext._preds_AICcObs and (B) values from the most probable model, which are obtained from the UCODE_2005 output file for model 3D, with path and filename data-win\obsonly\z3\4\z._linp.

Model-averaged intervals tend to be larger than intervals calculated for single models. Comparing the individual model-averaged confidence intervals shown in Figure A2-5A with those shown for model 3D in Figure A2-5B yields the following. The model averaged intervals are, on average 62 percent wider, but the values for each prediction range widely. The model-averaged interval is actually slightly smaller for prediction o22off (98 percent the width of the interval for model 3D), and more than four times larger for prediction o25off.

It is important to investigate the calculations behind model-averaged values. One thing of interest is which models dominate the model-averaged values, and whether it is most useful to report the model-averaged predictions or predictions from the one or two most probable models. Figure A2-6 shows the rankings of the models considered in the example problem. The model with rank 1 is the most probable model. Many of the models have the same large rank of 11, indicating that they have been omitted from consideration. The reasons for omitting models from consideration by MMA are discussed at the beginning of chapter 2 of this report. The values plotted in Figure A2-6 are from the MMA output file testmma_ext._rank

The values of the AICcObs model criterion are shown in Figure A2-7. The rankings of Figure A2-6 are directly related to these values: smaller AICcObs values result in smaller rankings, which indicate better models. In Figure A2-7, omitted models are excluded and the values plotted are from the MMA output file testmma_ext._mma.

The posterior model probabilities and the inverted evidence ratios expressed as a percent are shown in Figure A2-8. Table A2-2 lists the model analysis statistics calculated using the AICcObs model criterion, which is the AICc criterion calculated only using observations. In this problem, prior information is not defined. These results suggest that based on this model criterion, model 3D is more probable than the other models given the observations. The values plotted in Figure A2-8 are from the MMA output file testmma_ext._anals_AICcObs.

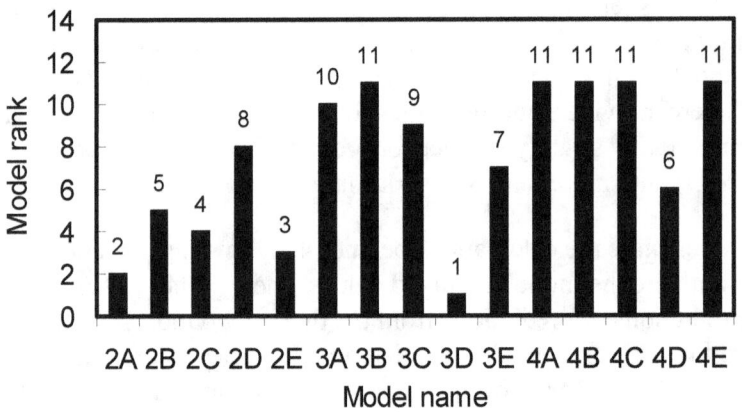

Figure A2-6. Model ranks based on the AICcObs model criterion. Smaller values identify better models. The repeated largest value of 11 indicates models excluded from the analysis.

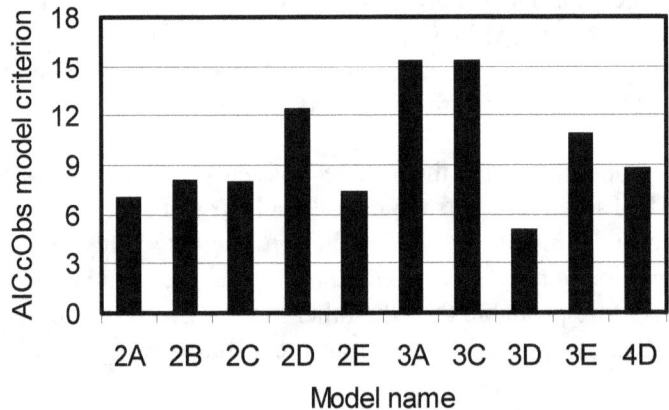

Figure A2-7. Values of the AICcObs model criterion. Smaller values identify better models.

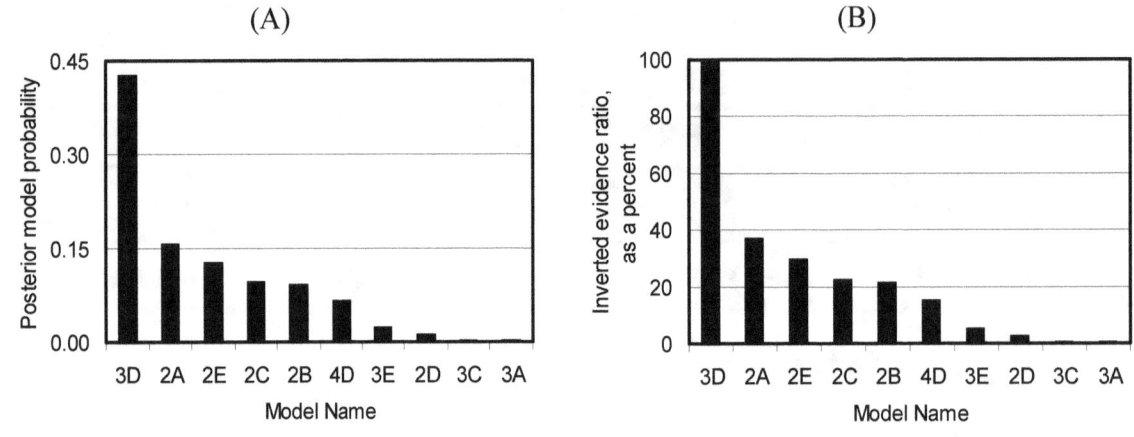

Figure A2-8. (A) Posterior model probabilities and (B) inverted evidence ratios expressed as a percentage for the models included in the MMA analysis, calculated using the AICcObs model criterion. For both, larger values identify better models.

Table A2-2. The AICcObs model criterion values, and resulting delta values, posterior model probabilities, evidence ratios and inverted evidence ratios for the example problem. The values are from the MMA output file testmma_ext._anals_AICcObs.

[AICcObs, the AICc criterion of equation 2.2a evaluated for observations only, as described in chapter 3; eq., equation number of this report]

Model	AICcObs	Delta value (eq. 2.3)	Posterior model probability (eq. 2.4)	Evidence ratio (eq. 2.5)	Inverted evidence ratio, as a percent (eq. 2.6)
2A	6.97	1.98	0.16	2.69	37.16
2B	8.07	3.08	0.09	4.67	21.40
2C	7.96	2.97	0.10	4.41	22.65
2D	12.37	7.38	0.01	40.09	2.49
2E	7.42	2.43	0.13	3.37	29.66
3A	15.33	10.34	0.00	175.86	0.57
3C	15.29	10.30	0.00	172.70	0.58
3D	4.99	0.00	0.43	1.00	100.00
3E	10.88	5.90	0.02	19.07	5.24
4D	8.75	3.76	0.06	6.56	15.24

In chapter 2, in the discussion after table 1, it is suggested that the delta values (Δ_i) can be used as follows: models with $\Delta_i < 2$ are very good models; models with $4 < \Delta_i < 7$ have less empirical support and, in most cases, models with $\Delta_i >$ about 10 can be dismissed from further consideration. Here, model 3D is the best model, but model 2A has considerable support, and models 2E, 2C, 2B, and 4D have notable support. According to this analysis, models 3A and 3C could be dismissed from further consideration. Given the similarity of all aspects of the models except the distribution of hydraulic conductivity, the poor fit of the lower ranked models is likely due to their poor representation of the connections among and locations of hydraulic conductivity zones. Poor representations include, for example, more or fewer high hydraulic conductivity paths through the system.

Finally, it is important to consider the model fit to observations and results of sensitivity analysis of the most probable models. Figure A2-9 shows the Cook's D values calculated for model 3D. Cook's D is a measure of influence, as discussed, for example, by Hill and Tiedeman (2007). The critical value is calculated as 4/ND, where here ND=21. Larger values of Cook's D identify observations that are more important to the estimated parameter values. Here, observations o13 and o15 exceed the critical value. The values plotted are from the model output file data-win\obsonly\z3\4\z. rc.

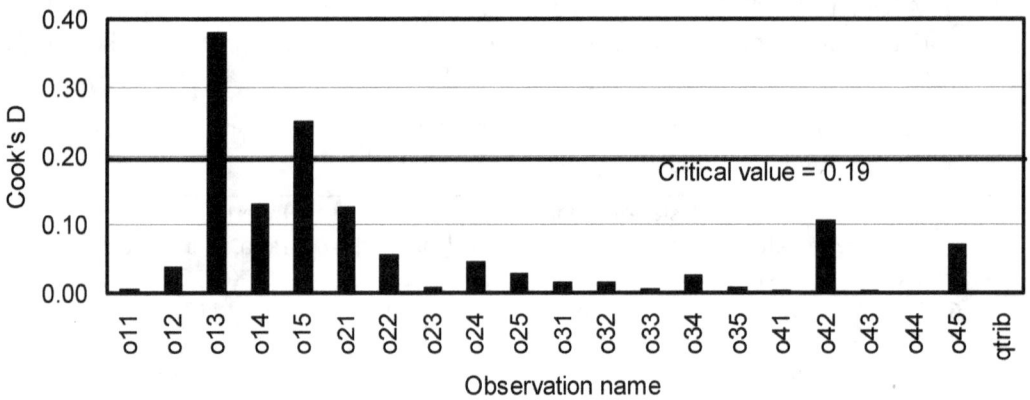

Figure A2-9. Cook's D for the model named 3D, which is top-ranked from the analysis of models with observations only using the AICcObs model criterion.

The results shown in Figure A2-9 suggest that head observations o13 and o15 are the most important observations for the estimated parameters values. Looking at Figure A2-3, using the observation location names listed in Table A2-1, it can be seen that the most important head observations occur along the west-most line of observation locations. It is not obvious why these observations are most important, and it is likely to be related to the hydraulic conductivity distribution. In ground-water models, flow observations like qtrib often are important. However, in this system calibration conditions include substantial imposed pumpage and qtrib is not an important observation to parameters according to the Cook's D measure. Observations that are not important to parameter estimates can be important to predictions. This can be measured using the OPR statistic that can be calculated using the OPR-PPR computer program (Tonkin and others, in press).

Appendix 3: Distributed Files and Directories

MMA files are distributed in the directories listed in Table A3-1.

Table A3-1. Directories distributed with MMA.
[MMA, multi-model analysis; GHz, gigahertz; CPU, central processing unit; RAM, random access memory]

Directory	Subdirectories	Description
data-*os*[1]	**obsonly** (contains subdirectories **z2**, **z3**, and **z4**, which each contain subdirectories **1**, **2**, **3**, **4**, and **5**) **Prior** (Contains subdirectory **z2**, which contains subdirectories **1**, **2**, **3**, **4**, and **5**)	Each subdirectory obsonly/z2/1, prior/z2/3, and so on contains the input and output files for regression, prediction, and confidence intervals for one model. Model names consist of the number following the "z" and a letter between A and E that corresponds sequentially to the 1 to 5 used as directory names. For example, the model in subdirectory z2/4 is named 2D. To remove the output files in these directories, use the batch file CleanRegressions.bat in the subdirectory or 02-Clean_Regressions of the test-*os* directory. To re-generate the output files use the batch files in subdirectory 03-Run_Regressions of the Test-*os* directory.[2]
test-*os*[1]	**01-Run_MMA**	Contains three batch files and associated MMA input files. One set of files produces a minimal analysis, one produces an extensive analysis, and the third produces an analysis for models that include prior information equations. There is also a batch file to remove the output files produced when MMA is run.
		The following files are not needed to execute the MMA test cases, but provide an example of how to set up models in preparation for evaluation using MMA.
	ReadMeFirst.txt	Instructions for running the models used by the test cases.
	02-Clean_Regressions	A batch file to delete the regression output files in the subdirectories of the data-*os* directory.
	03-Run_Regressions	Contains one batch file for each model considered. Running one batch file conducts a regression, simulates predictions, and calculates confidence intervals on predictions and parameters for one model. The input and output files for the runs are in the data-*os* directory.
doc		The pdf file for this document.
bin		Executables used in the example simulations.
src	mma api-modules ucode-modules	Fortran files unique to the MMA program. JUPITER API modules used by MMA. UCODE_2005 modules used by MMA The source for the other executables included in the bin directory can be obtained through the Web.[3]

111

[1] The content of this directory is operating-system dependent. The directory name is formed by substituting letters representing the operating system for "*os.*" For example, test-win includes files for a Windows operating system.

[2] For most computers using a Windows operating system, the models can be run conveniently using Windows Explorer by highlighting any number of batch files, right clicking the mouse, and choosing 'open' and OK. However, some computers cannot handle this many executions at once, and the computer will terminate. Therefore, save all other work before attempting to run many batch files at once for the first time. If multiple batch files are executed simultaneously, each will open a run in a separate window. The command windows will close when the executions are complete. The run takes about 5 minutes on a Pentium 4, 3.2 GHz CPU with 1GB of RAM. Once the runs are complete the files included in the original distributions will be restored and MMA can be used to evaluate the results.

[3] http://water.usgs.gov/nrp/gwsoftware/modflow2000/modflow2000.html and http://water.usgs.gov/software/ucode.html

Appendix 4: Discrepancies between the Equation for KIC Presented in this Work and those Presented Elsewhere.

The second term of equation 2.8 presented here for KIC differs from that used by Carrera and Neuman (1986). The difference occurs because the Fisher information matrix (Fisher, 1922), which equals

$$\text{Fisher information matrix} = (1/\sigma^2)\,\underline{X}^{\mathrm{T}}\,\underline{\omega}\,\underline{X}, \tag{A4.1}$$

needs to be divided by n to normalize for use in calculating KIC. When the determinant is expanded and terms are combined, equation 2.8 results.

The equation for KIC used by Poeter and Anderson (2005) has the same deficiency. In addition, in Poeter and Anderson (2005), the Fisher information matrix was mistakenly defined as $\underline{X}^{\mathrm{T}}\,\underline{\omega}\,\underline{X}$ instead of using equation A4.1. Using the Fisher information matrix results in the first term of equation 2.8 being correctly multiplied by $(n-(k-1))$ instead of the incorrect n used by Poeter and Anderson (2005), where $k-1$=NPE is the number of parameters estimated for the process model and k and NPE are defined after equation 2.2a.

www.ingramcontent.com/pod-product-compliance
Lightning Source LLC
Chambersburg PA
CBHW081459170526
45166CB00008B/2489